AF540985

Conservation Agriculture and Ecology

About the Authors

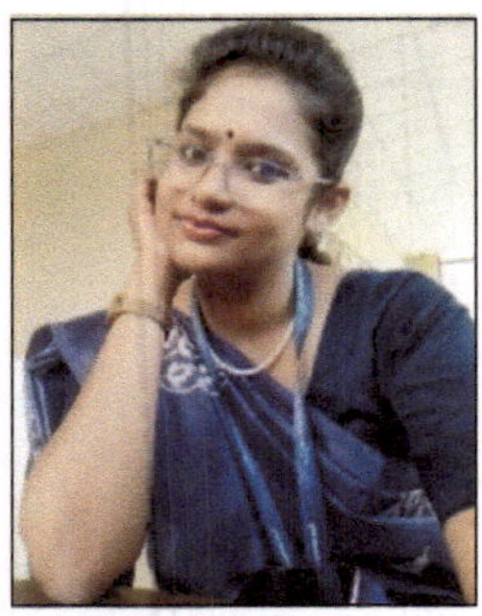

Dr. Anwesha Mandal is currently working as an Assistant Professor in School of Agricultural Sciences, G D Goenka University, Haryana, India. She did her PhD in Agricultural Extension from BCKV, the premier agriculture University India under the guidance of Prof, S K Acharya. Miss Anwesha Mandal has completed her graduation from Palli Siksha Bhavana, Visva Bharati (2016) in Agriculture and post graduated from G B Pant University of Agriculture and Technology, Pantnagar, Uttarakhand (2018) with ICAR-JRF (NTS) in Agricultural Extension and Communication. Thereafter, she joined Bidhan Chandra Krishi Viswavidyalaya in the department of Agricultural Extension to complete her PhD as a Senior Research Fellow under Center for Advanced Agricultural Science and Technology (CAAST) on Conservation Agriculture Project funded by ICAR-NAHEP and World Bank. She has cleared both UGC-NET (Nov., 2017) in Adult Education/Non formal Education/ Andragogy and ICAR- NET (2018) in Agricultural extension in her maiden attempt. She was a Visiting Fellow in Department of Natural Resources, CALS, Cornell University, Ithaca, New York, USA for 6 months during her PhD programme. She is the Best PhD Thesis Awardee, 2023, from the National Institute of Agricultural Extension Management (MANAGE), Hyderabad, India Her research area is mostly focused on conservation and management of natural resources through raising awareness and sensitizing the farmers. She has presented her research work in both national and international platforms and also got recognition as Best Oral Presenter. She has also authored and co-authored a few books, book chapters and good no. of research papers in some reputed journals.

Prof. (Dr.) Sankar Kr Acharya, presently, Dean, Post Graduate Studies; former Head, Department of Agril Extension and Director, Extension Education, Bidhan Chandra Krishi Viswavidyalaya, Mohanpur, WB, born on 6th October, 1960, started his career as Assistant Professor at BCKV in 1988 and has been in teaching, research and extension over 35 years. An erudite teacher as well as an elegant speaker, is internationally acclaimed for his unique research domain of Social Entropy and Energy Metabolism, Social Ecology and Environmental Sociology, Enterprise Ecology Framework, Conservation Stewardships: Research Publication: 247 in National and International Journals. Book Publication: 113 books authored. Expert Member, Agriculture Commission; Marketing and Extension Sub Committee, Govt. of West Bengal; Expert member WWF projects in BTR (Buxa Tiger Reserve Project); Expert member; DFID project on Primary Education (DPEP); Rapid Environment Impact Analysis, Visited Italy, France, Germany, China , Sri Lanka and Bangladesh. He has also been honoured to be selected as the Convener of Panel (PE-32) entitled The hunger, poverty and silence, of IUAES, University of Manchester, UK, 2013 .Co-PI of ICAR, NAHEP and World Bank funded project on Conservation Agriculture. So, far he has successfully guided 21 Ph D scholars,. He is the fellow of ISEE, IARI, New Delhi, Fellow, BIOVED, He is now acting as editor, reviewer of a good number of national and international journals. Fellow Award, ISEE(ICAR-IARI, New Delhi), 2019 Awards and Distinctions: Distinguish Scholar Award, Krishisanskriti, New Delhi; Certificate of Merit for standing First class first at M Sc(Ag) in Agricultural Extension, 1986; Honorary Appointment to the Research Board of Advisors, The American Biographic Institute (2003) ;He has been honored to be selected as the Convener of Panel (PE-32) entitled the hunger, poverty and silence, of World Congress, IUAES, University of Manchester, UK, 2013; He has been honored to be selected as the Convener of Panel (P-102) entitled,' Uncertainties, Unpredictability and Marginalization: The impact and mitigation for survival of agriculture and humanity, of the 19th World Congress, IUAES-WAU, 14-20 October, 2023, New Delhi, India, 2023; Honored to be nominated as Expert Member, Working group on Agricultural Extension, West Bengal State Agriculture Commission (2007); Honored to be nominated by the Hon'ble Vice-Chancellor, BCKV to Act as an Expert Member, DPEP, A DFID, UK, project for planning and revamping primary education through POA of GOVT of India.(1995) Expert Member, Buxa Tiger Reserve Project funded by WWF(World Wide Wild Life Funding), 1998; Dr Daulat Singh Memorial Award, Society of Extension Education, Distinguished Professor and Academician Award, Research Education Solution (RES), M S Swaminathan School of Agriculture, Centurion University (2022); Achievers of German patent (Utility model no.20202310273) :Novel LOT based inventory management system with radio controlled pallet racking for storage solution. He is the Fellow, West Bengal Academy of Science and Technology (WAST). He has also been recipient of the Fellow, Mobilization Award, IARI, New Delhi.

Conservation Agriculture and Ecology
Evolving Perception at Farmers Level

Anwesha Mandal
Assistant Professor
School of Agricultural Sciences
GD Goenka University, Haryana

Sankar Kr Acharya
Dean, Post Graduate Studies
Former Head, Department of Agril Extension
Director Extension Education
Bidhan Chandra Krishi Viswavidyalaya
Mohanpur, West Bengal

FU 75, Pitam Pura
New Delhi 110034
Mob. 91-8447075807
Email: publishmywork23@gmail.com
Web: www.publish-mywork.com

Print ISBN: 978-93-61342-48-6

ebook ISBN: 978-93-61345-33-3

Preface

The global agriculture is now reeling under serious threats and, these are emanating from the persuasion of a terribly wrong agriculture over a protractile period resulting into serious disruption of ecology and its cognate functional impact. The problem has become more complex due to unpredictable pace of global warming and climate change. India is to support a mammoth population size of 1600 million by 2050 with less and lesser land and water resources, and more and more unpredictable weather incidences. Around 68 per cent of land mass of India is suffering from the acute organic carbon depletion and, this has got a serious and appalling impact on crop productivity. Water availability to defend agricultural production process is dwindling very fast. During 1950s, per capita availability of water had been 500 cubic meter per year, which has been dwindled to only 1700 cubic meter/capita/yr. Every hectare of India's crop land is being eroded of 10-15 tons of top soils due to heavy tillage and intense farm operations including heavy tillage and soil disruption. These all bleak informations have hit our mind and logic to go for conservation agriculture. The book has hugely focused on the trend of conservation agriculture and its evolving perceptions in farmers' way of taking and confronting agriculture from the perspectives of ecology and economy. The book has sucked the sap of well structured research on conservation agriculture and its incoming execution and responsibility as perceived by practicing farmers of a two characteristically different agro-climatic zones of Bengal , namely, the terai and new alluvial zones. The book , as I expect and trust , shall generate an appeal to the global audience comprising of scholars , faculties and policy makers.

Authors

Contents

1

Introduction

1.1 Background

The aftermath of Green Revolution has contributed to both coercions and pollution, albeit, it saved human civilization from hunger, malnutrition and impoverishment. The beginning of GR era, with the prevailing objective and condition, has got less but few choices to go for intensive cultivation that invited chemical fertilizer, assured irrigation and farm mechanizations to make the 'magic gene' revolutionizing the production, productivity and livelihood security in agriculture as well.

The welcome of exotic genes into Indian agriculture has also triggered the silent of traditional cultivars. Nevertheless, the quantum jump of an agriculture with a production of 55 million tons to 120 million tons during 70s, has contributed to make India self-reliant in food and we have been able to earn a new esteem in the global geo-political scenario by throwing away the 'begging bowl' called PL480.

The euphoria of GR starts evading with the erosion of soil, contamination of ground water, narrowing down of gene base and decline of ecological resilience. We are now producing more than 300 million tons of foodgrains, but at the same time we are concerned on the plateauing of production and productivity scenario of certain crops including rice and wheat. We must say, not much of progress has already been reflected in areas of coarse grain except maize. In any case, we have to scale up our total foodgrain production to 350 million tons by 2030, which sounds a herculean task with a fact that soil erosion has been further aggravated and water depletion to support agriculture almost remains unabated, when the fact stands that almost 50% of Indian agriculture is supported by groundwater.

The decline of soil organic carbon, which provides the fulcrum for all kinds of mineralization process as well as stand of microorganism count in the soil regime has posed a serious threat to the overall ecological resilience in the domain of agriculture and allied sectors. The decline of soil organic carbon

(SOC) level has gone congenital with the degradation of livestock population. When the count of livestock is being withdrawn from the agro-ecosystem, the burning of rice and wheat stubbles, harshly disproportionate application of NPK and apparent as well as ruthless stoicism to the silent erosion of local germplasms, the problem remains complex and challenging.

Associative with the depletion of groundwater (GW), there has been issues like thermal balances and global warming. If the GW depletion keeps maintain its present magnitude, by 2050, India will turn into a water stress nation. The social ecology of water stress involves both conflict and chaos, not to mention its severe effect on agriculture alone.

Till 90s, our farmers have been mostly trained on what would be the required amount of NPK etc. per hectare, leaving them apparently ignorant about the importance of balanced nutrition and its correlates with production behavior of crop and resilience of surrounding ecosystem. In addition, the highly soluble nitrogen are happily welcomed by GW due to its smartest leaching losses to invite another problem called 'blue baby syndromes' through the intake of toxic water with nitrates. Also, indiscriminate use of pesticides are seamlessly contributing to the entries of Hg, Cd, Cu into the food chain to have a forward journey with huge biological magnifications. The increase of carcinoma, cardiac failures, neural stresses and fatigued are being contributed to those who are in direct contact, mostly farmers who ensures our food security.

So, the need of the hour is to change the popular rhetoric of food security. The volume of contaminated food can go no longer ensuring security of health and mind. We need to produce food with zero level toxicity, but at the same time, it should be sufficient to supplement our calorie, protein and mineral requirements.

Production of safe food and good agricultural practices cannot be possible without conservation agriculture. We cannot go ecologically prodigal by burning crop residues and killing microorganisms with our scripted knowledge and retrenched concepts. Farmers' participation and involvement is utmost required in ensuring least soil disturbances by phase wise attrition of number of tillage and by diverting the green residues into an integral part of soil compositions to make it well resilient and productive. The exuberance of flood irrigation cannot be allowed by letting a pumpset oozing out precious GW beyond recommendations and scales. So water literacy should be an inevitable and required knowledge amongst the farmers.

Crop diversification can be well effective only when it would accommodate more of legume crops like pulses and water saving crop like mustard etc.

The productive function of OC, soil moisture appropriateness in the soil, belligerence of microorganisms functioning, application of well scaled fertilizer combination and with increasing non-reliance on fossil fuels as source of energy can make our agro-ecosystem well-scripted for an orchestrations integrating ecology, economy and sociology together.

The social ecology of conservation agriculture (CA) invites casting deep introspection into the functional cohesiveness of ecology, economy and sociology. The market responsiveness to CA produces steers on the collateral processes like soil health management, crop diversification, irrigation socialization and shimmering perception of CA consequent. The rationalization of the tillage operation must convey the economic benefits which a farmer can be accrued to, at the same time, the lesser disruption of the soil should carry a proportionate ecological dividend to the farmers through an experiential learning cycle.

The experience based learning as proposed by Kolbs, starts with experience. So, the dividend of CA should be in the form of real **experiences** and these experience, may be in the form of increased income and quality of CA, shall lead to a processing or a reflection that is the farmer should be able to correlate between experience and the surrounding ecological resilience. After **processing** the patterning of socialization, CA will be reflected in an ever increasing application and follow up of conservation principles. Application germinates theory, the prime CA principles and philosophy, a changed, informed and inculcated persuades of knowledge and skills will make the farmers an ecology stewards rather than a robot for production. That's why, this research work is envisaging into an interactive yet complex domain of social ecology in the realm of CA. So, the inquiries are expected to be constructed, organized and enacted persuades of this research to put up the questions like

i. Are farmers aware of and concerned to the degradation of organic carbon, may be, in terms of their earned and local connotations?

ii. Do they really feel or change their water use behavior in compliance with or in response to the principles of the rhetoric conservation agriculture?

iii. What are the factors inter-alia conditions are making them fertilizer literate yet nutrient illiterate?

iv. Has ecosystem management got any perceptual inputs in the psychic construct of the farmers that could be a drivers for present and future ecosystem management in CA?

v. Are the predicted variables, perception on water management, nutrient management and carbon management correlated and interactive with the set of predictor variables tried out under the study?

The social ecology of ownership and holding pattern and its seamless fragmentation is sure to make our farm energy prodigal and economically extravagant. The process of fragmentation and its geo-spatial detachments from each other in the form of land ownership and proprietorship has got a long term effect on resource, energy and cost management in our tiny farms. The practices of livestock regime as domestic economy, which has got a rich tradition, now losing its ground as a pride and profit for rural livelihoods. The regime of livestock has unfortunately inherited a subdued occupational identity in this part of Bengal and the concept of integrated farming is increasingly becoming more theoretical and less empirical. The burning of crop residues to add to this misery has somehow earned a psycho-managerial support in favor of its apparent economic dividend. What little CR could have been added to the soil, has been perished in the form of smoke to add to the rate and intensity of pollution of global warming. In the absence of strong policies to curb the practice of stubble burning has made these pernicious process remain unabated. Market has got the indomitable contribution towards the initiation, growth and evolution, as per expectation and direction, in making CA a meaningful ways and means to millions of farmers surviving with fragmented yet tiny holdings.

Disintegration of joint families into small and scattered nuclear families has becoming isochronous phenomenon related to fragmentation and erosion of land holding capabilities. So the question may shimmer; can there be a community approach to CA else it has to face an infertile consequences for many more previous experiments we have had so far. Social networks among community plays a strong role in disseminating information on access to CA technologies. Enhanced institutional linkages between research, extension and farmers aids in timely decision-making on when and what farm operations are needed to be carried out for improved agricultural production and productivity (Chisanga et al., 2017).

1.2 General Objective

To study the socio-ecological factors influencing the perceptional change of the farmers on water, nutrient and carbon management in conservation agriculture.

1.3 Specific Objectives

The present study has been set up with the following specific objectives:

i. To study the present status of perception and practices of Conservation Agriculture in terms of water-nutrient-carbon management among the farmers

ii. To identify and standardize the set of variables in operation, impacting on the behavior of water, nutrient and carbon management of the farmers;

iii. To elucidate and interpret the complex interactions amongst and between sets of variables, as predicted and predictor characters, as a system function and policy implication in CA;

iv. To organize series of participatory analyses and action learning to generate a dynamic repository of knowledge covering both indigenous and conventional approaches.

2

Citation and Reviews

This chapter describes about the review of relevant researches related to the present study. This also helps in developing conceptual framework and selecting proper methodology for the study. Although studies have been conducted on different aspects of conservation agriculture in past, but a few reviews were available on social ecology and perceptional change of the farmers in managing critical inputs to agriculture like water, nutrient and soil organic carbon, and ecosystem as a whole. So, reviews having a little indirect bearings are also included in this section. In accordance with the subject of the proposed study, the review of literature has been presented under the following sub-heads:

2.1 Socio- Economic and Communication Characteristics of The Farmers

Gholve (1986) in his study on socio-cultural and psychological correlates of tribal development stated that majority of the respondents (62.10%) were having small land holdings (upto 2ha), followed by 20.53 percent and 17.37 percent of the respondents having medium and big size of land holdings (2.1 ha-4.0 ha and > 4ha) respectively.

Sakharkar (1995) in a study on knowledge and fertilizer use pattern among soyabean cultivating farmers of Nagpur in Maharashtra found that 36.00 per cent of the respondents had participated in one or more extension activities whereas, the rest of the respondents did not participate in any extension activity.

Reddy and Rao (1998) carried out a study on assessing knowledge, attitude and adoption of farmers about improved practices in agriculture reported that majority of the farmers were illiterate (64.67%), owned small size farm (54.67%) and nearly 40 per cent of them had moderate extension contact.

Kanavi (2000) in a study on knowledge and adoption behavior of sugarcane growers in Tamil Nadu reported that none of the respondents participated regularly in training and demonstrations. Nearly one-third (31.33%) of the respondents participated in Krishimela. Very less number of respondents (1.33%) participated in extension activities like farm visits, group discussion (2.66%) and study tours (4%).

Raghunandan (2004) found that the major percentage of the respondents (73.75%) were literate. Maximum of the respondents (22.50%) had attended primary school, followed by high school (15%), pre university (11.25%) and graduation (5%), whereas, 23.28 per cent of the respondents were illiterate.

Thiranjangowda (2005) in a study observed that, the majority (40.62%) of the respondents belonged to high farming experience category, followed by medium experience (35.93%) and low experience category (23.45%).

Rayanagoudar (2009) in a study revealed that most of the respondents in organic village were of middle aged category (53.00%), illiterates (61.00%), had large land holdings (61.00%), high social participation (90.00%) and television was the most popular mass media owned by 66.66 per cent of them.

Tatlidil et al. (2009) in a study found that the higher the socioeconomic status (more frequent contact with extension services, higher education, ownership of land, etc.) and the greater the access to information, the greater the perceived importance of sustainable agricultural practices.

Singh, et al. (2010) in a study on relative influence of socio-personal characteristics and utilization of fertilizer technology in wheat cultivation highlighted that majority (63.33%) of the small farmers were in 31 to 45 year age group and living in pucca houses, having nuclear families, educated up to high school (87.15%) with an annual income range from 10,000 to 25,000 and were member of atleast one organization.

Oyesola and Obabire (2011) in a study reported that farmers in the study area were mostly male with a mean age of 53.8 years, married, and had formal education. Crops grown by the farmers include: maize, yam, cassava, plantain, vegetables, and tomato. Farmers' sources of information on organic farming were radio, extension agents, television, newspapers, farmers association, fellow farmers, and relatives. Their most preferred sources of information were Mobile phones and radio.

Badgujjar (2012) found that the majority of respondents (42.50%) had medium farming experience followed by high 33.75 per cent and low farming experience 23.75 per cent.

Dutta (2012) reported that majority of the farmers practicing farm diversification belonged to middle aged group (67.00%) with education up to primary level (30.00%) and were having nuclear type of family (64.00%). Majority of these farmers were marginal farmers (67.00%) with medium family income (66.00%), medium farming experience (68.00%), medium level innovativeness (60.00%), medium scientific orientation (54.00%), medium cropping intensity (72.00%) and no social participation (70.00%).

Meena (2012) in a study on knowledge level of FFS participants revealed that majority of the respondents were in middle age (54.55%), had agriculture as main occupation (543.64%) and practiced mixed farming (85.45%). The study also found majority of the participants were having medium farming experience (68.18%), medium level of innovativeness (52.72%), medium extension agency contact (61.82%) and medium mass media exposure (80.00%).

Mohapatra (2013) in a study on adoption of soil health management practices found that majority of the respondents were in middle aged group (60.00%), had medium level of extension participation (72.22%), were members of gram panchayat (68.89%), belonged to medium socio-economic status (53.34%) and had moderate level of mass media utilization (68.88%).

Patel (2013) in a study conducted on green manuring perceived that majority of the farmer respondents belonged to middle aged group (56.67%), had secondary to higher secondary level of education (55.00%) with above 15 years of farming experience (56.67%), medium level of mass media exposure (50.00%) and low level of extension contact (46.67%). It also revealed that most of the respondents falls under small farmer category depending on the size of land holding (51.67%), had an annual income of more than 2,00,000 rupees (35.00%) and a high level of scientific orientation (51.67%).

Suryawanshi, et al. (2017) in a study conducted on the vermicompost trainees found that out of the total trainees, 50.00 per cent belonged to middle age and higher percentage (40.52%) were middle school passed. Majority of these trainees were from small farmers category (60.34%), had medium social participation (58.62%), low income (24001/- to 84000/- per annum), marginal land holding (up to 1.00 ha) (44.82%) and were using the vermicompost in their field. It was also found that majority (75.86%) of the vermicompost trainees were having agriculture as their main occupation.

Naik, et al. (2018) in a study on knowledge level of farmers on organic red gram cultivation practices reported that half of the respondents (50%) were young aged. Equal percentage of respondents (54.20%) were having small farm size with low farming experience and medium socio-economic status. More than half of the respondents (55%) were reported to have moderate risk orientation.

Tapoopi et al. (2018) in their study reported that in case of dissemination of CA information among the farmers, NGOs played the major role followed by service providers who mostly owned farm implements and farmer to farmer exchange of knowledge. Although agricultural extension officers and media also played a role in CA information dissemination, their impact was outweighed by NGOs and service providers.

2.2 Perception of Farmers on Nutrient Management in Conservation Agriculture

Ramamurthy (1976) also found that farm size had a significant positive influence upon IR-8 adopters and their perception of profitability in India due to their greater access to information sources and more knowledge in farming. Farmers' education, farming experience and communication exposure also had significant positive influence on farmers' perception.

Ervin and Ervin (1982) found that education has a positive impact on the adoption of soil conservation technology. They also found that older farmers are less likely to use soil conservation practices, whereas younger farmers may be better educated and involved with more innovative farming.

Shortle and Miranowaski (1986) found experience has a positive effect on the adoption of conservation tillage practices in the Four Mile Creek Watershed of eastern Iowa.

Moore (1988) concluded that according to teacher educators, a myriad of factors, such as social stratification, political instability, transportation, and national education policies appear to be the responsible for the agricultural and educational development problems.

Bruening, Radhakrishma, and Rollins (1992) maintained that farmers who had more than a high school education perceived water pollution, manure mismanagement, and nutrient mismanagement as more serious environmental issues than those farmers who had not completed high school.

Williams and Wise (1997) reported that agricultural teachers perceived themselves as having additional things to learn about sustainable agriculture practices and students rated themselves as only knowing a little about them.

Kashem and Mikuni (1998) did not find any significant influence of farm size and farmers' education upon their perception of indigenous technical knowledge (ITK) in Japan.

Sheikh et al. (2003) revealed that in the 'cotton–wheat' system personal characteristics like education, tenancy status, attitude towards risk implied in the use of new technologies and contact with extension agents are the main factors that affect adoption. As regards the 'rice–wheat' system, resource endowments such as farm size, access to a 'no-tillage' drill, clayey soils and the area sown to the rice–wheat sequence along with tenancy and contact with extension agents were dominant in explaining adoption.

Ahmed and Al-Kadi (2004) argued that higher perception ratings were observed for those who utilize more sources of information and preferred group extension.

Duncan (2004) reported that the secondary educators either agreed or strongly agreed that the agricultural training program will contribute to students' success in the agriculture industry and that the program offers a valuable educational experience for students.

Farouque and Takeya (2007) reported that different categories of farmers indicated that landless, marginal and small farmers had a low level of awareness when compared with medium and large farm holders. The overall perception of farmers in the study areas revealed that a significant proportion (78%) had either a low or a very low level of perception while 22% had a medium to high level of perception. Findings from individual interviews with farmers indicated that they perceived themselves as having a low perception of preparation of farm yard manure and the role of organic matter as well as the beneficial aspect of ISF and NM for sustainable crop production. Among the nine characteristics of farmers, four characteristics- education level, farming experience, farm size and communication exposure influenced farmers positively. However, two characteristics- family size and fertilizer use negatively influenced farmers' perception of ISF and NM for sustainable crop production.

Fusun Tatlidil et al. (2009) pointed out that farmers who have higher socio-economic status, i.e. higher education, ownership of land, more frequent contact to extension services and better access to information, tend to have greater perceived importance of sustainable agricultural practices.

Chisanga et al. (2017) in their results revealed that the major hindrances to the application of CA practices lay in biophysical, technological, land, institutional and agro-climatic constraints. Generally, farmers reported weeds as a major biophysical constraint to the implementation of CA technologies

Daxini et al. (2018) in a study found that for the national sample (n=1009) attitudes, subjective norms (social pressure), perceived behavioural control (ease/difficulty) and perceived resources are significant and positively associated with farmers' intentions. In case of the voluntary sample (n=587), only attitude, perceived behavioural control and perceived resources are significantly and positively associated with farmers' intentions. Whereas, for the mandatory sample (n=422), subjective norms, perceived behavioural control and perceived resources are significantly and correlated in a positive direction with intentions. A number of farm and farmer characteristics are also significantly associated with intentions.

Ntshangase et al. (2018) in their study deciphered that farmers' positive perceptions were positively correlated with higher maize yields. Their study also revealed that an increase in extension visits, age, education and farmers' positive perceptions significantly increased the likelihood of a farmer adopting no-till CA, but an increase in land size was negatively related to no-till CA adoption. Their findings confirmed the important role of extension in the promotion of no-till CA, particularly the intensity of the extension services.

Tapoopi et al. (2018) in a study found that technological know-how, limited agricultural inputs and implements for conservation agriculture hindered the uptake of conservation agriculture. In addition, lack of crop residues for mulching purposes and little understanding of the importance of crop rotation were identified as barriers to practice conservation agriculture.

Mugandani and Mafongoya (2019) in their study found that although fewer than 10% of the respondents had received any formal training in agriculture, more than 80% of them relied on it for their major source of income. The respondents had high levels of knowledge on the social, environmental and economic benefits of conservation agriculture. However, the majority of the non-adopters had an indifferent perception towards conservation agriculture. The knowledge and perception of the farmers was explained by age, gender, education and experience with conservation agriculture.

Lopes et al. (2020) found that farmers who perceived residue burning as diminishing soil quality are far less likely to do so; however, their awareness of its adverse environmental effects does not lower the choice to burn. They also inferred that farmer's wealth increases the likelihood of residue burning. Identifying the behavioral factors behind residue burning, which capture underlying aspects of social influence and herd behavior was found to be useful for policy formulation aiming to reduce this public health hazard.

2.3 Ecological Resilience and Conservation Agriculture

Peterson, et al. (1998) in their study described the existing models of the connection between species diversity and ecological function, and proposed a conceptual model that relates species richness, ecological resilience, and scale. They suggested that species interact with scale-dependent sets of ecological structures and processes that determine functional opportunities.

Adger (2000) defined social resilience as the ability of groups or communities to cope with external stresses and disturbances as a result of social, political and environmental change. The origins of this interdisciplinary study in human ecology, ecological economics and rural sociology are reviewed, and a study on the impacts of ecological change on a resource-dependent community in contemporary coastal Vietnam in terms of the resilience of its institutions has been outlined.

Gunderson (2000) in his study concluded how ecological resilience is key to management of complex systems of people and nature.

Walker, et al. (2004) in a study based on three related attributes of social–ecological systems (SESs) to determine their future trajectories, interpreted the SES dynamics for sustainability science include changing the focus from seeking optimal states and the determinants of maximum sustainable yield (the MSY paradigm), to resilience analysis, adaptive resource management, and adaptive governance (process of creating adaptability and transformability in SESs).

King (2008) explored how agri-ecological systems contribute to more sustainable and resilient communities, through community development processes and concluded that ecological system theories and concepts might contribute to thinking about the future of community based agri-ecological resilience.

Shava, et al. (2010) looked into how agricultural knowledges may be adapted and applied among relocated people in light of globalizing trends toward urbanization and resettlement. The study revealed that the communities draw upon their reserves of knowledge to respond to changes within their local environments. Such knowledge can serve as a source of community resilience by enabling people to sustain their livelihoods and community well-being, and thus adapt to environmental changes and displacement.

Dumas, et al. (2016) in their study found that highly variable rainfall, lack of education, agricultural inputs and market access constrained many farmers of Zambia's Luangwa Valley to employ short-term coping strategies that threaten ecosystem resilience. Community Markets for Conservation (COMACO) promotes household, agricultural and ecological resilience along two strategic lines: improving recovery from shocks (mitigation) and reducing the risk of shock occurrence.

Acharya and Chatterjee (2021) in a study found that CA is gaining more attention of the farming community throughout the world. They also reported that in comparison to 2008-09, 69% more area was brought under CA in 2015-2016.

2.4 Ecosystem Services and Conservation Agriculture

Nelson, et al. (2009) observed that there is an increasing consensus about the importance of incorporating "ecosystem services" into resource management decisions, but quantifying the levels and values of these services was proven difficult. They used a spatially explicit modeling tool, Integrated Valuation of Ecosystem Services and Tradeoffs (InVEST) to predict changes in ecosystem services, biodiversity conservation, and commodity production levels.

Power (2010) emphasized on agricultural ecosystems which rely on ecosystem services provided by natural ecosystems, including pollination, biological pest control, maintenance of soil structure and fertility, nutrient cycling and hydrological services. The study focused on the tradeoffs that may occur between provisioning services and other ecosystem services and disservices and should be evaluated in terms of spatial scale, temporal scale and reversibility. As more effective methods for valuing ecosystem services become available, the potential for 'win–win' scenarios increases.

Sanderson, et al. (2013) explored the concept of diversity and its role in maximizing ecosystem services from managed grasslands and integrated agricultural systems (i.e., integrated crop–livestock–forage systems) at the field and farm level. The study revealed that integrating dynamic cropping systems with livestock production increases the complexity of management, but also creates synergies among system components that may improve resilience and sustainability while fulfilling multiple ecosystem functions.

Farooq and Siddique (2014) found that CA involves changing many conventional farming practices as well as the mind-set of farmers to overcome the conventional use of tillage operations.

Palma, et al. (2014) revealed that Conservation agriculture (CA) changes soil properties and processes compared to conventional agriculture. These changes can, in turn, affect the delivery of ecosystem services, including climate regulation through carbon sequestration and greenhouse gas emissions, and regulation and provision of water through soil physical, chemical and biological properties. Conservation agriculture can also affect the underlying biodiversity that supports many ecosystem services.

Giller, et al. (2015) found that over the past 10 years CA has been promoted among smallholder farmers in the (sub-) tropics, often with disappointing results. Growing evidence challenges, the claims that CA increases crop yields and builds-up soil carbon although increased stability of crop yields in dry climates is evident. Their analyses suggested a pragmatic adoption on larger mechanized farms, and limited uptake of CA by smallholder farmers in developing countries. They have also proposed a rigorous, context-sensitive approach based on Systems Agronomy to analyse and explore sustainable intensification options, including the potential of CA.

Garbach, et al. (2016) provided an illustrative examination of the framework, evaluating evidence for yield and ecosystem service outcomes across five AEI systems: conservation agriculture, holistic grazing management, organic agriculture, precision agriculture, and system of rice intensification (SRI). The review presents substantial evidence that the five AEI systems can contribute to multifunctional agriculture by increasing ecosystem service provision, or reducing negative externalities associated with agriculture, while maintaining or increasing yields.

Ghosh, et al. (2019) explained how Conservation agriculture (CA) has emerged as a promising technology for efficient use of available resources and sustained productivity in the long run by saving inputs, reducing energy use and greenhouse gases emissions through CA-based management practices. The major focus was on the potential ecosystem service benefits accrued from CA. The paper concluded that CA could be a strategy for sustainable crop intensification and a climate resilient crop management system in the long run.

Acharya and Chatterjee (2019) reported in our operating agro-ecosystem, water is contaminated, soil is eroded, organic carbon is depleted and biodiversity is marginalized. Conservation Agriculture aims to protect and preserve the precious natural resources, stabilizing and augmenting food production and livelihood for millions.

2.5 Stewardship, Ecology and Conservation Agriculture

Worrell and Appleby (2000) explained stewardship as a potentially useful concept in modernizing management philosophies. The connections between the practical usage and the ethical basis of stewardship are currently poorly developed. They looked stewardship as the responsible use (including conservation) of natural resources in a way that takes full and balanced account of the interests of society, future generations, and other species, as well as of private needs, and accepts significant answerability to society.

Foster, et al. (2003) demonstrated the ubiquity and importance of land-use legacies to environmental science and management and recognizes that the legacies of land-use activities continue to influence ecosystem structure and function for decades or centuries—or even longer—after those activities have ceased. Consequently, recognition of these historical legacies adds explanatory power to the understanding of modern conditions at scales from organisms to the globe and reduces missteps in anticipating or managing for future conditions. As a result, environmental history emerges as an integral part of ecological science and conservation planning.

Chapin, et al. (2010) in a study revealed that ecosystem stewardship is an action-oriented framework intended to foster the social–ecological sustainability of a rapidly changing planet. Recent developments identify three strategies that make optimal use of current understanding in an environment of inevitable uncertainty and abrupt change: reducing the magnitude of and exposure and sensitivity to known stresses; focusing on proactive policies that shape change; and avoiding or escaping unsustainable social–ecological traps.

Hails and Ormerod (2013) observed that national and international assessments are increasingly highlighting the unsustainable use of earth's natural resources in the face of population increase, growing material affluence and global change. The concept of natural capital emphasizes on managing the environment as an asset, ensuring that the benefits are sustained for future generations.

Mathevet, et al. (2018) explored and discussed the various meanings of the stewardship concept in the field of sustainability science. This study presented two dimensions for mapping the diversity of stances concerning stewardship and also analysed these positions in relation to the limits of the systemic approach, ideological manipulation, responsibility, and solidarity. The study concluded how the concept of ecological solidarity can contribute to the underpinning of a specific form of social-ecological stewardship.

2.6 Erosion of Natural Resources and Issues of Ecological Resilience

Whitlow (1988) discussed the main factors influencing land degradation are population density and farming systems that dominate the physical risks of erosion. Conservation history in commercial and peasant farming areas is reviewed as a basis for understanding the current status of erosion and a future conservation strategy was outlined in terms of political, educational, agrarian and research issues.

Knickel (1990) observed that current farm practices contribute to a wide array of environmental problems like monoculture, soil erosion, contamination of ground water resources by nitrate and pesticides, wildlife habitats loss or subject to fragmentation and highlighted that environmental policies need to be an integral part of regional development and agricultural policies.

Wilkinson (2011) made a theoretical contribution to the emerging inter-disciplinary exploration by engaging questions like what, if any, new conceptual ground does social-ecological resilience offer planning theory, and more broadly what issues does social-ecological resilience raise for further scholarship by planning theorists.

Nichols, et al. (2011) observed that climate change and its associated uncertainties are of concern to natural resource managers. Their study encompasses on adaptive resource management which is an application of structured decision-making for recurrent decision problems with uncertainty, focusing on management objectives and the reduction of uncertainty over time.

Cross, et al. (2012) presented a model for collaborative planning aimed at identifying ways to adapt management actions to address the effects of climate change in landscapes that cross public and private jurisdictional boundaries.

Nearing, et al. (2017) observed that the regions of land that are brought into crop production from native vegetation typically undergo a period of soil erosion instability, and long term erosion rates are greater than for natural lands as long as the land continues being used for crop production. Broadly applied soil conservation practices, and in particular conservation tillage and no-till cropping, were found to be effective in reducing rates of erosion with the widespread adoption of new conservation tillage and residue management practices.

2.7 Water Conservation and Conservation Agriculture

Andersson and D'Souza (2013) concluded a more thorough analysis of farming households and their resource allocation strategies is required to understand the farm-level adoption constraints different types of farmers face. As contextual factors appear key influences on smallholders' farming practices, studies focusing on the wider market, institutional and policy context which are needed to understand (limited) CA adoption in Southern Africa.

Kassam, et al. (2014) overviewed achievements in soil and water conservation on agricultural lands through the experience derived from the adoption and spread of Conservation Agriculture (CA) world-wide.

Maggioni (2015) found that mandates to curb outdoor water uses are correlated with reductions in residential per capita water usage, while water rates and subsidies for water saving devices are not. It also confirms that size is a significant policy implementation factor. Three factors could improve the conservation effort: using prices as a conservation tool, not only as a cost recovering instrument; investing in water efficient tools only when they provide significant water savings; supporting smaller agencies in order to give them opportunities to implement conservation strategies more effectively or to help them consolidate.

Lal (2015) observed strategies how to use CA judiciously and prudently to harness its potential as a conservation-effective technology, climate-resilient agriculture, and a viable option for sustainable intensification of agroecosystems for advancing food and nutritional security, and for adaptation/mitigation of climate change.

Belay, et al. (2020) improved our understanding of water and nutrient dynamics in green pepper grown under CA and CT. Use of CA provides opportunities to optimize water use by decreasing irrigation water requirements and optimize nutrient use by decreasing nutrient losses through the runoff and leaching.

2.8 Soil Conservation and Conservation Agriculture

Hurni, et al. (2008) studied that small-scale farming, the largest occupation in the world, is highly affected by soil erosion along with its alternative environmental threats by decreasing yields which were primarily used for subsistence. Therefore developing soil and water conservation measures are necessary to safeguard small-scale farming, which needs external support for investment in sustainable land management technologies as an imperative and integral element of farm activities.

Thierfelder and Wall (2010) conducted a study showing clear advantages of CA practices with residue retention and crop rotation over conventionally tilled practices in terms of higher populations of earthworms in CA plots, higher total carbon and more water stable aggregates and better infiltration rates.

Busari, et al. (2015) reported many edges of conservation tillage over standard tillage (CT) with reference to soil physical, chemical and biological properties similarly as crop yields. Conservation tillage involving ZT and minimum tillage that has potential to interrupt the surface compact zone in soil with reduced soil disturbance offers to steer to a more robust soil surroundings and crop yield with least impact on the surroundings.

Du, et al. (2021) aimed to plan SOC thickness (0–30 cm) utilizing cross breed strategies that consolidate territorial topographic models and a remaining kriging methodology utilizing local samples from an area of interest and assess the impacts of various sampling designs for residual kriging on final map quality. This examination exhibits the utility of extrapolating a territorial topographic model by utilization of a limited number of new examples for local alignment and gives understanding to specialists who are keen on mapping soil carbon distribution in sparsely sampled areas of interest.

2.9 Biodiversity and Conservation Agriculture

Dumanski, et al. (2006) studied the relationships between conservation agriculture and zero tillage. Application of CA promotes the construct of optimizing yields and profits whereas making certain provision of native and world environmental advantages and services. Zero tillage, along with other soil conservation practices, is the cornerstone of CA.

Mcneely and Schroth (2006) highlighted the considerable potential of traditional agroforestry practices to support biodiversity conservation. These embody the importance of ample areas of natural surroundings and of appropriate hunting regulations for maintaining high levels of biodiversity in agroforestry land use mosaics, as well as the critical role of markets for tree products and of a favorable policy environment for agroforestry land uses.

Kumaraswamy and Kunte (2013) focused on the major challenge to achieve food security and meet the demands of the ever-growing human populations while simultaneously conserving biodiversity, providing critical ecosystem services and maintaining rural livelihoods. The findings suggested to implement offset mechanisms, which are compensatory and balancing approaches to restore the ecological health and function of an ecosystem to mitigate the devastating situation and integrate biodiversity with agricultural landscapes.

2.10 Barrier to Conservation Agriculture

Knowler and Bradshaw (2007) conducted a study on farmers' adoption of conservation agriculture. The study concludes that efforts to promote conservation agriculture will have to be tailored to reflect the particular conditions of individual locales.

Greiner and Gregg (2011) exhibited the benefit of planning research in a way that empowers esteem added investigation so that, on one hand, the consequences of contextual analyses can be approved and summed up and, on the other hand, local contrasts can be elicited and adequately considered to be in policy design and program execution.

Perry-Hill and Prokopy (2014) analysed the distinctions in awareness, attitudes, constraints, and behaviour of agricultural, small agricultural and rural residential landowners. They found that small agricultural and rural residential landowners have more uplifting perspectives, yet are less acquainted with water quality issues and protection associations. Additionally, a small agricultural or rural residential landowner is related with more prominent readiness to attempt conservation practices. For all landowners, perceived cost was the greatest obstruction to conservation practice adoption.

Pittelkow, et al. (2014) concluded that any expansion of conservation agriculture should be done with caution, as execution of the other two standards is frequently difficult in asset poor and weak smallholder cultivating frameworks, thereby improving the probability of yield losses instead of gains. The investigation from this examination shows that the potential contribution of no-till to the sustainable intensification of agriculture is more restricted than frequently accepted.

Mwase, et al. (2015) evaluated a few publications on selection of agroforestry in Southern Africa and complemented the review with household and key source meetings to get proof from farmers and promoters of the advancements on the variables influencing adoption.

2.11 Extraction of Aroundwater and its Impact on Ecological Resilience

Katic and Grafton (2011) assessed the trade-offs among resilience and economic payoffs in terms of groundwater extraction where there is a danger of an irreversible and cataclysmic occasion. The outcomes show that assuming the limit is dubious, controlling both the rate and profundity of extraction can create a higher monetary return and a lower likelihood of crossing the threshold than just controlling the rate of extraction.

Arfanuzzaman and Atiq (2017) found that population growth has a large influence on water demand to rise. The study proposed an integrated SWDM approach, which incorporates optimum pricing, ground and surface water regulation, water conservation, sustainable water consumption and less water foot print to ease groundwater depletion.

Grönwall and OduroKwarteng (2017) found that water insecurity is a developing concern worldwide, particularly for developing countries. The role of groundwater and springs in buffering the impacts of environment fluctuation is progressively recognized, yet it must be completely acknowledged with a more powerful understanding of groundwater as a resource, and how utilization of it and reliance on it contrast. Conjunctive use, managed aquifer recharge, and suitable treatment measures are vital to make groundwater an essential asset on the metropolitan plan.

2.12 Soil Erosion and Its Impact on Ecological Resilience

Lal (1997) focused on global soil degradation, its extent and agronomic impact, which can only be resolved through understanding of the processes and factors leading to establishment of the cause-effect relationships for major soils, ecoregions and land uses. Efficient assessment through long haul experimentation is required for setting up quantitative measures in relation to the ease of restoration through judicious management and discriminate use of essential input.

Cumming, et al. (2005) studied about the idea of social-ecological resilience holds guarantee for interdisciplinary amalgamations. They have equated resilience with the capacity of a system to keep up with its identity, where system identity is characterized as a property of key segments and connections (organizations) and their progression through space and time. This methodology will likewise yield experiences into the instruments of progress and the expected results of various approach and the management decisions, giving a degree of choice help for each contextual analysis region.

Kosmas, et al. (2016) explored the socio-environmental framework dominated by pasture land in Asteroussia Mountains (Crete, Greece) between1950 and 2010 with the intend to distinguish changes in the most significant framework's credits, as a commitment to the investigation of land debasement according to the flexibility viewpoint. The grazing system was sustainable in the previous time span being agreeably dealt with tolerably low soil disintegration rates in spite of antagonistic soil, geographical and environment conditions.

Teague, et al. (2016) suggested that with suitable regenerative crop and grazing management, ruminants diminish generally GHG outflows, yet in addition facilitate provision of essential ecosystem services, increment soil carbon (C) sequestration, and decrease ecological harm. The paper infers that to guarantee long-term sustainability and ecological resilience of agroecosystems, agricultural production ought to be directed by arrangements and regenerative administration conventions that incorporate ruminant grazing.

Flores, et al. (2019) led an investigation on tropical woods which are threatened by increasing natural and anthropogenic aggravation systems. In view of a broad writing survey, the specialists proposed an applied model in which soil disintegration builds up unsettling influence impacts on tropical woods, diminishing their resilience with time and improving their probability of being trapped in an alternative vegetation that is vulnerable to erosion.

2.13 Biodiversity and Its Impact on Ecological Resilience

Oliver, et al. (2015) focused on resilience as opposed to transient conveyance of ecosystem functions and services, and the thought of specific underpinning mechanisms, which has assisted with joining the research areas of biodiversity–ecosystem function and ecological resilience and finally aid the development of evidence-based yet flexible ecosystem management.

Winn and Pogutz (2013) presented key ideas from ecology and social ecology to organization and management studies-ecosystems, biodiversity, ecosystem services, and ecological resilience. The researchers represent these ideas with progresses in ecosystems management and conclude with ideas for future examination in sustainability management, organization theory, and strategic management.

Jackson, et al. (2010) conducted a study about the quick changes in land use, food systems, and livelihoods require social-ecological systems that keep various choices open and get ready for future unpredictability. With three models, the utilization and conservation of agrobiodiversity is investigated along temporal, spatial, and human institutional scales for its role in sustainability: first, farmers' seed systems; second, complex pollination systems; and third, wildlife conservation in agricultural regions with high destitution. Incentives are necessary if agrobiodiversity is to give advantages to people in the future.

Di Falco and Chavas (2008) investigated the powerful impacts of rainfall shocks on agroecosystems productivity. It reports the adverse impacts of decrease in rainfall on the agroecosystem productivity both in the short run and long run. The experimental proof shows how higher diversity upholds resilience and keeps up with the system productivity under challenging climatic conditions.

Moller, et al. (2008) led an examination where an obvious proof has been shown that agricultural intensification has degraded amphibian biodiversity. The examination gathers a summed up probability that intensification has likewise decreased terrestrial biodiversity and agro-ecosystem resilience. This survey closes with a crucial statement: a shift of emphasis is now desperately required inside New Zealand land-use management, research and policy.

2.14 Awareness, Knowledge and Action for the Restoration of Ecosystems

Su, et al. (2021) revealed that no tillage (NT) is often presented as a means to grow crops with positive environmental externalities, such as enhanced carbon sequestration, improved soil quality, reduced soil erosion, and increased biodiversity. The study expanded existing datasets to include the results of the most recent field experiments, and produce a global dataset comparing the crop yields obtained under CT and NT systems. The dataset can help to gain insight into the main drivers explaining the variability of the productivity of NT and the consequence of its adoption on crop yields.

Chapin, et al. (2015) presented a stewardship framework for a more practical direction for arctic species, ecosystems, and peoples—to some extent expanding on promising existing endeavors, and somewhere else supporting new segments. The framework more thoroughly incorporates social and ecological elements of conservation across scales and anticipates and shapes multi-scale changes. Stewardship consequently gives an integrated framework for conservation practice, including useful direction for society to further develop conditions for conservation in a quickly evolving world.

Ballard and Jill (2010) inspected a Participatory action research (PAR) and tracked down that the PAR approach did surely create environmental learning, characterized here as ecological literacy, civic literacy, values awareness and self-efficacy and added to resiliency through advancing more diversity, memory, redundancy, and adaptive capacity. The investigation inferred that the PAR approach is an important tool for environmental learning however the degree to which learning can really promote system change and greater resilience should likewise be perceived inside the basic setting, particularly political realities.

King, (2008) conducted a study on current research which explore a range of agri-ecological systems and include systems such as Organic Agriculture, Biodynamics, Community Supported Agriculture (CSA), Permaculture, Farmers Markets and Community Gardens and how they contribute to agri-ecological and community resilience. The study reviewed and concluded by comparing ecological systems models to agri-ecological systems, and suggests how ecological systems theories and concepts might contribute to thinking about the future of community-based agri-ecological resilience.

Fiksel, (2006) emphasized on a comprehensive system approach which is necessary for effective decision making with regard to global sustainability, since industrial, social, and ecological systems are firmly connected. Various exploration bunches are utilizing dynamic modelling techniques, including bio complexities, system dynamics, and thermodynamic analysis, to research the effects on ecological and human systems of significant moves, for example, climate change and the related policy and technology responses. These methods can yield something like a partial understanding of dynamic system behaviour, empowering a more coordinated way to deal with systems analysis, beneficial intervention, and improvement of resilience. Recommendations are provided for continued research to accomplish progress in the dynamic modeling and sustainable management of complex systems.

Füsun Tatlıdil, F., Boz, İ. and Tatlidil, H. (2009). Farmers' perception of sustainable agriculture and its determinants: a case study in Kahramanmaras province of Turkey. *Environment, Development and Sustainability,* **11. pp:** 1091–1106. https://doi.org/10.1007/s10668-008-9168-x

depending on prioritising existing enclaves and [illegible] or [illegible] supporting new systems. The [illegible] [illegible] [illegible] and ecology [illegible] [illegible] across scales and [illegible] patches and shapes, multi-scale change [illegible] gives a [illegible] framework for conservation practice, [illegible] useful direction [illegible] society to further develop conditions for [illegible] in the [illegible] evolving world.

Ballard and [illegible] (2010) [illegible] a participatory action research (PAR) and [illegible] down that the PAR approach [illegible] environmental learning, [illegible] ecological literacy, [illegible] and self-efficacy and [illegible] to resiliency [illegible] memory, [illegible] and adaptive capacity. The [illegible] inferred that the PAR approach is an important tool for [illegible] learning [illegible] the degree to which learning [illegible] change and [illegible] resilience should likewise [illegible] particularly political realities.

King (2008) completed a study [illegible] of agro-ecological systems [illegible] such as [illegible] Agriculture, Biodynamics, Community Supported Agriculture (CSA), Permaculture, Farmers Markets and Community Gardens [illegible] agro-ecological [illegible]. The study [illegible] concluded by comparing [illegible] and suggests how ecological systems [illegible] thinking about the future of sustainable agroecological [illegible].

[illegible] (2010) [illegible] a comprehensive systems approach which is necessary for effective decision-making with regard to global sustainability since industrial, social and ecological systems are [illegible] interconnected. Various [illegible] and [illegible] and modeling techniques, including [illegible], system dynamics and thermodynamic analysis, to assess the effects of ecological and [illegible] systems. For example, climate change [illegible] and [illegible] methods can [illegible] something like a [illegible] understanding and [illegible] system behaviour, [illegible] systems analysis, behavioural interventions and improvement in resilience. Recommendations are provided for continued research to accomplish progress in the [illegible] and sustainable management of complex systems.

[illegible] (2009) [illegible] perception of sustainable agriculture and [illegible] [illegible] provinces of Turkey. *[illegible]*, [illegible] 109–[illegible]. [illegible]

3

Theoretical Orientation

The last century has witnessed a global change in agricultural practices. From high dependency on external input agriculture, the world is now shifting towards low external input sustainable agriculture (LEISA). The history of dustbowl devastation in 1930s in America has put agricultural practices, tillage in particular, to question (Friedrich et al., 2012). Traditionally, tillage was believed to be a primary farm operation for preparing a good seedbed, nutrient mobilization, weed management, moisture availability, crop residue incorporation (Hobbs et al., 2008), thereby improving the overall soil health with increased production (Farooq et al., 2011a). The dust bowl devastation of 1930s gave rise to the concept of protecting soil by reducing tillage operation and maintaining ground cover, popularized as conservation tillage (Friedrich et al., 2012). Economic and ecological sufferings caused by the disastrous droughts in USA during 1930s drove the shift towards conservation agriculture (CA) (Haggblade and Tembo, 2003). However, it took more than 20 years for CA to reach its desired adoption in South America (Friedrich et al. 2012) and only during 1960s no-tillage could make an entry into the farm practices in USA. During this time, farm equipment and management practices for no-tillage systems were tested, trialed, improved and developed to optimize the crop performance and machinery and farm operations (Friedrich et al., 2012). The positive experiences then helped promoting CA in other parts like South Africa, Brazil, Paraguay etc. (Haggblade and Tembo, 2003). The technology spread exponentially in the early 1990s with more farms adopting CA practices in countries like Brazil, Argentina and Paraguay (Friedrich et al., 2012). The increasing popularity and quick adoption of the technology grabbed the attention of many national and international research organization like FAO, CIMMYT, CIRAD and some CGIAR centers. These organizations showed their keen interest in promoting the conservation led farming practices and played a major role in adoption of CA farming systems in Africa and some parts of Asia. Presently, 180 mha are under CA practices worldwide, USA (43.2 mha) having the highest area under CA followed by Brazil (32 mha) and Argentina (31mha). In India, the spread is still inconsistent with only 1.5mha (Kassam et al. 2018).

Conservation agriculture is a resource-saving agricultural production system that aims to attain production intensification and competitive yields while enhancing the natural resource base. This is achieved in compliance to three linked principles of minimum tillage, permanent cover and diversified crop rotation, implemented with locally formulated good production practices including crop, nutrient, water and pest management practices (FAO, 2019). Besides conserving the natural resources, biodiversity and labor, CA also increases soil water availability, enhance water infiltration, tolerance to abiotic stress (viz. drought and heat) and betters soil health in long term (Colmenero et al. 2013, see website https://www.cimmyt.org/news/what-is-conservation-agriculture). CA principles help mitigating soil nutrient depletion, land degradation and also increase yields that structure the entire concept of Conservation Agriculture (Hobbs et al. 2008).

Conservation agriculture (CA) and climate smart agriculture (CSA) though sometimes used interchangeably, there lies a fine demarcation between the two. CA aims for sustainable intensification of small farms using natural processes having a positive effect on the environment, adapting and helping farmers to increase their profit in spite of the climate risk. On the other hand, CSA aims for adaptation and mitigation of negative effects of climate change through soil carbon sequestration and reducing greenhouse gas emission. It also ensures livelihood and food security of farmers by increasing system productivity and profitability under a changing climate. Thus, CA is a part of CSA, as it delivers its objectives.

Though India could achieve food security through Green Revolution, it led to over exploitation of natural resources coupled with indiscriminate use of inorganic fertilizers and pesticides, and thereby declining factor productivity, increasing soil salinity, loss of biodiversity, lowering of ground water table, environmental pollution, pest resurgence and land degradation are some of its consequences. Therefore, the advantages of the green revolution have now been masked by the problems posed by it. Though the Eastern region is rich in natural resources, its potential could not be harnessed in terms of improving agricultural productivity, poverty alleviation and livelihood improvement. Eastern region of India has been focused to user second Green Revolution so as to meet out the ever increasing demand of food in the country. However, it is possible only through improving the soil health, minimizing the impact of biotic stresses, increasing the water productivity, development of suitable varieties, and integrated approach of land use. Conservation agriculture, therefore, is need of the hour, particularly in Eastern IGP where rice-wheat cropping system is predominant.

Resource conservation technologies (RCTs) make use of natural resources more efficiently and save input for food production. Appropriate RCTs encompass innovative crop production systems that combine the objectives such as dramatic reductions in tillage with an ultimate goal to achieve zero till or controlled till seeding for all the crops in a cropping system if feasible, rational retention of adequate levels of crop residues on the soil surface to arrest run-off and control erosion, improve water infiltration and reduce evaporation, increase soil organic matter and other biological activity to enhance land and water productivity on sustainable basis, identification of suitable crop rotations in cropping system and crop diversification and intensification to boost food security, incomes and thereby provide the livelihood security to the people.

In a recent review paper on the practice of Conservation Agriculture (CA) (i.e., minimum mechanical soil disturbance, organic soil cover and crop species diversification) by smallholders [Johansen et al., 2012], the authors concluded that it was an approach that could improve soil health, decrease costs of production and increase crop profitability.

CA is a resource-saving agricultural production system that aims to achieve production intensification and high yields while enhancing the natural resource base through compliance with three interrelated principles, along with other good production practices of plant nutrition and pest management (Abrol and Sangar, 2006). Boatmann et al. (1999) pointed out that traditional agriculture results in land resource degradation, wildlife and biodiversity reduction, low energy efficiency and contribution to global warming problems; whereas, conservation agriculture stresses the very beneficial impacts of a conservative way of cultivation on the global environment (soil, air, water and biodiversity), compared to traditional agriculture (Derpsch et al., 2010; Derpsch et al., 2011). As per the definition of FAO, Conservation Agriculture is, "a farming system that promotes maintenance of a permanent soil cover, minimum soil disturbance and diversification of plant species, enhancing biodiversity and natural biological processes above and below the ground surface, which contribute to increased water and nutrient use efficiency and to improved and sustained crop production." CA promotes most soils to have a richer bioactivity and biodiversity, a better structure and cohesion, and a very high natural physical protection against weather (raindrops, wind, dry or wet periods). It also protects surface and ground water resources from pollution and also mitigates negative climate effects. Hence, CA provides excellent soil fertility and also saves money, time and fossil-fuel. It is an efficient alternative to traditional agriculture, attenuating its drawbacks.

As per FAO definition CA is to i) achieve acceptable profits, ii) high and sustained production levels, and iii) conserve the environment. It aims at reversing the process of degradation inherent to the conventional agricultural practices like intensive agriculture, burning/removal of crop residues. With CA, farming communities become providers of more healthy living environments for the wider community through reduced use of fossil fuels, pesticides, and other pollutants, and through conservation of environmental integrity and services. Conservation agriculture systems is a total paradigm shift from conventional agriculture with regard to management of crops, soil, water, nutrients, weeds, and farm machinery.

Conservation agriculture basically relies on 3 principles, which are linked and must be considered together for appropriate design, planning and implementation processes. These are:

1. ***Minimal mechanical soil disturbance:*** The soil biological activity produces very stable soil aggregates as well as various sizes of pores, allowing air and water infiltration. This process can be called "biological tillage" and it is not compatible with mechanical tillage. With mechanical soil disturbance, the biological soil structuring processes will disappear. Minimum soil disturbance provides/maintains optimum proportions of respiration gases in the rooting-zone, moderate organic matter oxidation, porosity for water movement, retention and release and limits the re-exposure of weed seeds and their germination (Kassam and Friedrich, 2009). Many researchers highlighted several benefits of no-till CA, amongst which are an improvement in soil fertility, labour savings, improved efficiency in water use, an increase in productivity and environmental sustainability (FAO, 2000; Lugandu, Enfors, 2009). Reduced tillage decreases the environmental footprint of agriculture, tillage costs, loss of soil organic matter and improves timeliness of crop establishment (Hobbs et al., 2008).

2. ***Permanent organic soil cover:*** A permanent soil cover is important to protect the soil against the deleterious effects of exposure to rain and sun; to provide the micro and macro organisms in the soil with a constant supply of "food"; and alter the microclimate in the soil for optimal growth and development of soil organisms, including plant roots. In turn it improves soil aggregation, soil biological activity and soil biodiversity and carbon sequestration (Ghosh et al., 2010). Maintaining a permanent cover protects the soil from direct raindrop impact (Busari, Kukal, Kaur, Bhatt, & Dulazi, 2015) as well as improves its organic matter content, water infiltration and storage capacity (Hobbs, Sayre, & Gupta, 2008).

3. ***Diversified crop rotations:*** The rotation of crops is not only necessary to offer a diverse "diet" to the soil micro-organisms, but also for exploring different soil layers for nutrients that have been leached to deeper layers that can be "recycled" by the crops in rotation. Furthermore, a diversity of crops in rotation leads to a diverse soil flora and fauna. Cropping sequence and rotations involving legumes helps in minimal rates of build-up of population of pest species, through life cycle disruption, biological nitrogen fixation, control of off-site pollution and enhancing biodiversity (Kassam and Friedrich, 2009; Dumanski et. al., 2006). Crop rotation and diversification improves dietary diversity (Pellegrini & Tasciotti, 2014), pest and disease control, biodiversity (Giller et al., 2015) and nutrient cycling (Kassam, Friedrich, Shaxson, & Bartz, 2014). The practice can also improve resilience to climate change (Jost et al., 2016).

3.1 Concept of Conservation Agriculture (CA)

The most popular definition of CA is "a concept of crop production to achieve a high and sustained level of production along with acceptable profit, while saving the resources, con- serving the environment as well" (FAO 2013a, b).

According to the Conservation Agriculture Group of Cornell University (2015), "Conservation Agriculture (CA) is a set of soil management practices that minimize the disruption of the soil structure, composition, and natural biodiversity. CA has proven potential to improve crop yields while improving the long-term environmental and financial sustainability of farming".

CIMMYT (Donovan 2020) described CA as a practice that "conserves natural resources, biodiversity, and labour. It increases available soil water, reduces heat and drought stress, and builds up soil health in the longer term".

CA can also be defined as the integration of ecological management with modern, scientific, agricultural production. It employs all modern technologies that enhance the quality and ecological integrity of the soil, but the application of these is tempered with the traditional knowledge of soil husbandry gained from generations of successful farmers. (Dumanski et al. 2006).

3.2 Social Dimensions to the Practice and Sustainability of CA

Though it is very challenging for a single study to cover all categories in detail, where each is an important aspect of the social dimensions of CA. In practice, these themes construct intersecting, interacting, and often interdependent areas (Bacon et al. 2012).

3.3 Socialization and Adoption behavior of the Farmers' Regarding Conservation Agricuture

Rogers (1962) proposed in his Diffusion of Innovations Theory that there are four main elements that influence the dissemination of a new idea, i.e. the innovation itself, communication channels, time, and a social system. There is a point within this rate of adoption at which an innovation reaches the critical mass.

In the Theory of Reasoned Action, human behaviour is mainly predicted by three cog- nitive components like attitudes, social norms, and intentions (Kuo et al. 2015). After the addition of perceived behavioural control (controlled by the resource availability, opportunities, skills, as well as perceived significance to achieve the end product, the previous theory was renamed as Theory of Planned Behavior (White et al. 2015). Where, Theory of Interpersonal Behaviour firstly put personal beliefs, attitudes and social factors related to the behaviour and secondly, how does it affect, people's cognition, social determinants and personal normative beliefs effect and thirdly possibility of performing a specific behaviour that is predicted by intentions, situations and past experience (Misbah et al. 2015).

3.4 Key Challenges in Conservation Agriculture

Conservation Agriculture is going to complete its centenary after 10 years. However, only 180 M ha of land is under CA, where the total cultivated land in the world is 1407 M ha. If we consider CA as an effective production system, then a question may arise: Why is CA not spreading more rapidly? Here are some of the possible challenges before the adoption of CA will be discussed.

Agronomic Challenges

Among agronomic challenges, limited experience of cover crops, crop residue–livestock conflicts, disease-pest survival, and control of competing vegetation (cost and labour requirements) are the major ones (Ares et al. 2015). Besides, due to competition between crop residues and livestock fodder, farmers should be provided with an alternate fodder source, sometimes cover crops are suggested (Hellin et al. 2013).

Farmers do not have adequate information on cover crops at their specific farm conditions. Sometimes farmers become afraid of carryover of pests and various pathogens from previous to next crop through the leftover residues (Pramanik et al. 2014). Maintaining mulch for years may lock up nitrogen, and it causes increased fertilization in the short term (IIRR and ACT 2005). The key challenges in the case of the third principle, i.e. crop rotation, farmers have hesitation to insert legumes within the crop rotation due to adjustment in the spacing (Baudron et al. 2007). There are also very few reliable markets for a number of leguminous crops and a shortage of improved legume seeds (Baudron et al. 2007; Haggblade and Tembo 2003). Though, Himmelstein et al. (2017) found no evidence that leguminous intercropping, reduced tillage, pesticides, or fertilizers treatments can cause an increase in yields and gross income of a farm.

Non-availability of appropriate equipment and machinery, and proper herbicides to manage weed and vegetation is also a notable cause of slow adoption of CA (Pramanik et al. 2014).

Socio-economic Challenges

For centuries, farmers believe tillage as a good farming practice and are taught as a must- to-do agricultural operation during land preparation for any crop. So, bringing change in their mindset is not possible overnight (Pandey 2016)

Insecure land tenure/farm size, unavailability of finance mechanisms, insufficient information regarding aversion of risk through CA, and misleading information on suitable farming tools/agronomic practices are some of the socio-economic challenges being faced by CA farmers (Ares et al. 2015).

The key challenges are lying within the need for a proper understanding of the CA principles, proper coordination, need for crop diversification for better health, crop rotation for breaking disease-pest cycles.

3.5 Status of Conservation Agriculture: Global as well as Indian scenario

As reported by FAO (2014), total area under conservation agriculture is around 157 million hectares, which is approximately 11% of the total cropland area. The major CA countries are USA (35.6mha), followed by Brazil (31.8 mha), Argentina (29.2 mha), Canada (18.3 mha) and Australia (17.7 mha). In India, it is still in the initial phases. Over the past few years, adoption of zero tillage and CA has expanded to cover about 1.5 million hectares (Jat et al., 2012; www.fao.org/ag/ca/6c.html). The major CA based technologies being adopted is zero-till (ZT) wheat in the rice-wheat (RW) system of the Indo-Gangetic plains (IGP). In addition to ZT, other concept of CA need to be infused in the system to further enhance and sustain the productivity as well as to tap new sources of growth in agricultural productivity. The CA adoption also offers avenues for much needed diversification through crop intensification, relay cropping of sugarcane, pulses, vegetables etc. as intercrop with wheat and maize and to intensify and diversify the Rice-Wheat system. The CA based resource conservation technologies (RCTs) also help in integrating crop, livestock, land and water management research in both low and high potential environments.

Despite more than 50 years of research and extension, adoption of CA is limited and piecemeal in India.

3.6 Water Management and Conservation Agriculture

Global water scarcity has always been in the headlines in the recent past. According to a report by WHO, within 2020 25major cities of the world will be under acute water stress, and India is not lingering far behind. Major cities like Bengaluru, Chennai are already into the crisis where a litre of water bottle costs even upto Rs 400! India, a country with 17.4% of total population of the world has access to only 4% of the total available freshwater on this earth. Agriculture sector alone consumes 70% out of 4% of freshwater. It has been calculated that to produce 1kilogram of rice 3000-5000 litre of water is required. Conservation agriculture practices like retention of ground cover helps in maintain the soil moisture for a longer period, thus reducing the irrigation frequency. Moreover, micro irrigation systems like drip irrigation and sprinkler irrigation helps in saving water upto 50-60%. Growing direct seeded rice has been found to save upto 70% of water in comparison to transplanted paddy. Rainwater harvesting and water recycling can also help in efficient water management for sustainable agriculture.

3.7 Nutrient Management in Conservation Agriculture

Derpsch (2007a) said, "Experience has shown that most things learned at university about fertilization and liming should be revised, and new concepts of fertility management for no-till systems need to be developed and applied. The main principle to keep in mind is that farmers should fertilize their soils rather than their crops." This is also heard being uttered as: "feed the soil and let the soil feed the plant." Being a biologically-based practice with an agro-ecological perspective, CA does not focus on a single commodity or species. Instead, it addresses the complex interactions of several crops to particular local conditions capitalizing on the complex systems of interactions involved when managing soil systems productively and sustainably.

In a fully established CA system the aim of fertilizer nutrient management is to maintain soil nutrient levels, replacing the losses resulting from the nutrients exported by the crops. Because CA systems have diverse crop mix including legumes, and nutrients are stored in the soil organic matter, nutrients and their cycles must be managed more at the system or crop mix level. Thus, fertilization would not anymore be strictly crop specific, with the exception of nitrogen top dressing (if required at all), but will be given to the soil system at the most convenient time during the crop rotation. With the management of legume crops, either as previous crop in the rotation or as component in a cover crop before the next cash crop "top" dressing with nitrogen can be replaced by the N captured by the legumes and released during the following cropping cycle at the required time (more legume content – earlier release, more grass content in cover crop, later release). Additionally, undisturbed soils are habitats for free living nitrogen fixing bacteria and there is rhizospheric fixation of nitrogen (Sprent and Sprent 1990).

There are many different ecological and socio-economic starting situations in which CA has been and is being introduced. They all impose their particular constraints as to how fast the transformation towards CA systems can occur. In the seasonally dry tropical and sub-tropical ecologies, particularly with resource poor small farmer in drought prone zones, CA systems will take longer to establish, and step-wise approaches to the introduction of CA practices seem to show promise (Mazvimavi and Twomlow 2006). These involve two components: the application of planting 'Zai-type' basins which concentrate limited nutrients and water resources to the plant, and the precision application of small or micro doses of nitrogen-based fertilizer. In the case of degraded land in wet or dry ecologies, special soil amendments and nutrient management practices are required to establish the initial conditions for soil health improvement and efficient nutrient management for agricultural production (Landers 2007). What seems to be important is that whichever pathway is followed to introduce CA practices, there is a need for a clear understanding of how the production systems concerned should operate as CA systems to sustain soil health and productive capacity, and how nutrient management interventions that may be proposed can contribute to the system effectiveness as a whole both in the short- and long-term *(20) (PDF) Nutrient Management in Conservation Agriculture: A Biologically-Based Approach to Sustainable Production Intensification.* Available from: https://www.researchgate.net/publication/228464075_Nutrient_Management_in_Conservation_Agriculture_A_Biologically-Based_Approach_to_Sustainable_Production_Intensification [accessed Jul 27 2021].

3.8 Conservation Agriculture and Soil Carbon Restoration

The Paris Agreement at the 21st Conference of Parties (COP21) of the United Nations Framework Convention on Climate Change (UNFCCC), set an agenda for reducing global warming below 2 degree C and limiting the temperature increase to 1.5 degree C by lowering GHG emissions to encourage climate resilience through diverse pathways without compromising food production. The "4 per 1000" initiative was launched as a part of the Lima-Paris Action Agenda promotes SOC sequestration to improve food security and mitigate climate change. To achieve this target, improved management practices should be adopted for C sequestration in agricultural, forest and wetland along with rehabilitation of degraded soils. Various institutions in more than 170 countries initiated a highly ambitious goal with the collaboration between scientists, educator and farmers, policymakers to implement suitable practices for increasing SOC stocks. In addition to this, 103 countries have set mitigation and adaptation targets related to agricultural practices. Minimum soil disturbance, a major principle of CA that permits maximum of 20-25% soil disturbance, helps in reduction in C loss as atmospheric C.

3.9 Social Ecology and Social Ecological systems

The term social ecology was first coined by Murray Bookchin. According to him, "social ecology is based on the conviction that nearly all our present ecological problems originate in deep seated social problems. These ecological problems cannot be understood, let alone solved, without a careful understanding of our existing societies and the irrationalities that dominate it. In order to solve their ecological and social problems, society should use collective wisdom, cultural achievements technological innovations, scientific knowledge and innate creativity." He also emphasized that understanding how human relates to each other as social beings is critical in addressing current and future ecological issues.

It is always said that 'changes takes time', and society is one important factor that promotes changes. A farmer's decision making is influenced by a number of factors like, socio-ecological factors, agronomic practices, economic factors, cultural factors, psychological factors, market accessibility, prevailing prices, peer group interactions etc. Social ecology is a continuous interaction between the bio-physical resource settings and human life process, continuous and evolving as well as conflicting and recycling in nature. These interactions between society and environment creates a social ecology for the farmer which reflects on his decision making process.

The concept of social-ecological system as a multidimensional system where natural and social spaces do not have a sharp and immobile border can be illustrated. This framework was proposed by Berks et al. (1998). In this flow diagram, ecosystem, people and technology, local knowledge and property rights have been shown to have bidirectional relationship with pattern of interactions, sustainable society and knowledge and entitled dynamics.

The social ecology as a system is supported by three basic things – physical complex, biological complex and social complex. While the physical complex is functionally attuned to two basic components – matter and energy, the biological complex is characterized with basic characters like genetics and metabolism. The social complex has been epitomized over two complexes and has been unique by two basic characters – intelligence and motivation.

The social ecology has been evolved and configured through a didactic transformation of both the physical and biological complexes, for example, access to water has become a social issue contributed by physical existence of water resources and its quality as determined by count of coliforms. Social ecology thus, has become an operational exposition of bio-dynamics as well as energy metabolism. So these two, energy metabolism and bio-dynamics i.e. flow of energy and its exponential configuration in the form of motivation and knowledge keeps nibbling along and across space of society to characterize the social ecology and its functionality. So, social ecology is basically a study of interaction and interrelated count of energy and character of energy flow governed by human drive and motivation and human intelligence that begets devices and innovation, defense and aggression, pre-creation and destruction.

3.10 Social System Theory: The New Age Extension Science

Systems theory is interdisciplinary theory about the nature of complex system in nature, society and science. System theory originated in 1920's to explain the interrelatedness of organisms in ecosystem, more specifically, it is a framework by which one can investigate and describe any group of objects that work in concurrency to produce some result. This could be a single organism, any organization or society or any information artifact.

Social system theory was developed by Neklas Luhman. This theory is an option for the theoretical foundation of human resource management. This theory determines system as machines and takes open system approach. It addresses non-linear system theory. It also defines social system as an autopoietic closed system.

3.11 Conceptual Framework for CA Practices

A theoretical framework that provides understanding of the theoretical relationships between important variables and the economic performance of the agricultural practices is depicted in Figure 1. Variables such as knowledge, information on agricultural practices and the attributes of the farmers may influence the adoption of CA practices. The perception on CA leads to either uptake or rejection of the CA practices. The adoption of CA practices may lead to increase in crop yields, increase in cost, labour savings, and improvement in soil fertility and diversification of livelihoods. On the other hand, the non-adoption of CA may lead to decrease in crop yields, decrease in profits, decline in soil fertility and non-diversification of livelihoods. A critical analysis about CA's ability to reverse the negative effects of conventional and traditional agricultural practices has for a long time been overlooked (Giller et al., 2009). Literature warns that if CA fails to materialize in tangible benefits that are at the fore of farmers' interest, such as increased food or income, prospects of its widespread adoption are low (Cary & Wilkinson, 1997).

Researchers reported that farmers practicing CA observed co-benefits in their field like improvement of soil structure, water filtration and soil moisture, easy thinning of crops and reduction in weed population (Tapoopi et al., 2018).

CA is one of sustainable intensification that is increasingly promoted by various international research centres, international non-governmental organizations (NGOs), faith based organizations and governments of southern Africa among others to overcome the problem of soil degradation, drought, low and unstable crop yields and high production costs. CA is defined by three principles which must be applied simultaneously (i) minimum soil disturbance (ii) permanent soil cover with previous year's crop residues and (iii) diversification of crop species in sequence and/or in association (FAO, 2013). CA takes away the unsustainable components of conventional agriculture such as tilling the soil, removing organic material and monoculture and includes all other principles of sound crop management. While efforts have endeavoured to implement all the three principles of CA, often one or two of these principles have been applied by smallholder farmers. Consequently, partial application of the principles of CA do not lead to the desired modification of various agro-ecological functions such as soil health benefits, increased crop productivity and sustainability.

3.12 A Framework for Understanding Change: Conservation and Stewardship

The world is undergoing unprecedented changes in many of the factors that determine both its fundamental properties and their influence on society. Throughout human history, people have interacted with and shaped ecosystems for social and economic development (**Turner et al. 1990**, **Redman 1999**, Jackson 2001, Diamond 2005). During the last 50 years, however, human activities have changed ecosystems more rapidly and extensively than at any comparable period of human history (**Steffen et al. 2004, Foley et al. 2005, MEA 2005d**; Plate 1). Earth's climate, for example, is now warmer than at any time in the last 500 (and probably the last 1,300) years (IPCC 2007a), in part because of atmospheric accumulation of carbon dioxide (CO2) released by the burning of fossil fuels (Fig.3.1). Agricultural development largely accounts for the accumulation of other trace gases that contribute to climate warming. As human population increases, in part due to improved disease prevention, the increased demand for food and natural resources has led to an expansion of agriculture, forestry, and other human activities, causing large-scale land-cover change and loss of habitats and biological diversity. About half the world's population now lives in cities and depends on connections with rural areas worldwide for food, water, and waste processing. In addition, increased human mobility is spreading plants, animals, diseases, industrial products, and cultural perspectives more rapidly than ever before. This increase in global mobility, coupled with increased connectivity through global markets and new forms of communication, links the world's economies and cultures, so decisions in one place often have international consequences.

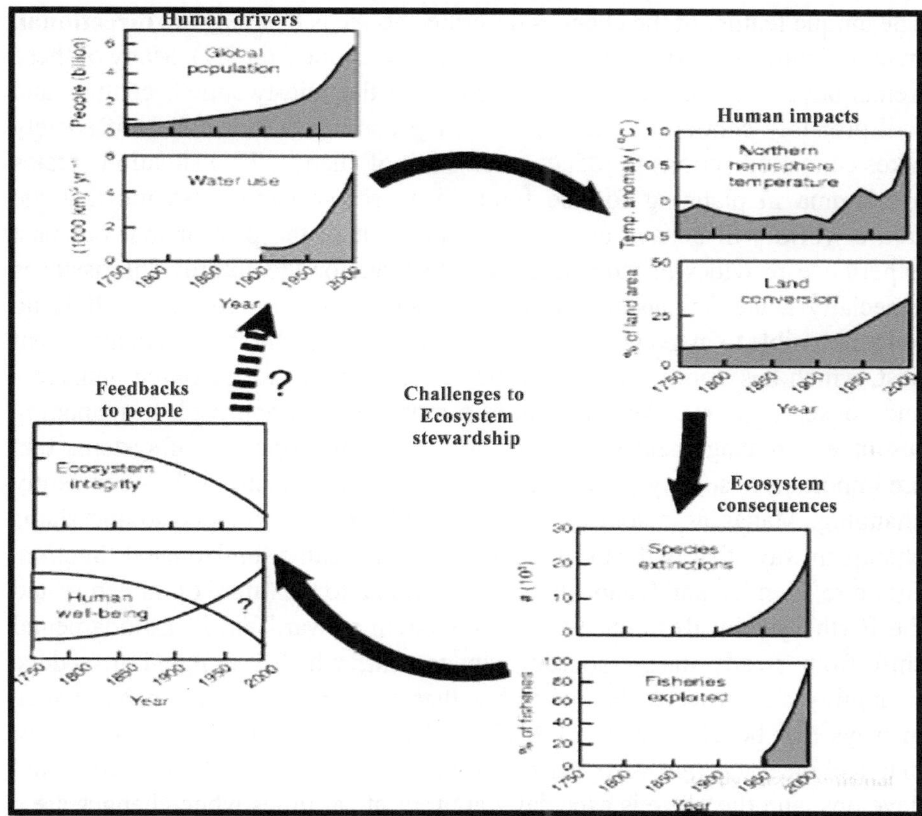

Fig. 3.1: Challenges to ecosystem

This **globalization** of economy, culture, and ecology is important because it modifies the **life-support system** of the planet (Odum 1989), i.e., the capacity of the planet to meet the needs of all organisms, including people. The dramatic increase in the extinction rate of species (100- to 1,000-fold in the last two centuries) indicates that global changes have been catastrophic for many species, although some species, especially invasive species and some disease organisms, have benefited and expanded their ranges. Human society has both benefited and suffered from global changes, with increased food production, increased income and living standards (in parts of the world), improved treatment of many diseases, and longer life expectancy being offset by deterioration in **ecosystem services**, the benefits that society receives from ecosystems. More than half of the ecosystem services on which society depends for survival and a good life have been degraded—not deliberately, but inadvertently as people seek to meet their material desires and needs (**MEA 2005d**). Change creates both challenges and opportunities.

The unique feature of the changes described above is that they are **directional**. In other words, they show a persistent trend over time (Fig.3.1). Many of these trends have become more pronounced since the mid-twentieth century and will probably continue or accelerate in the coming decades, even if society takes concerted actions to reduce some rates of change. This situation creates a dilemma in planning for the future because we cannot assume that the future world will behave as we have known it in the past or that our past experience provides an adequate basis to plan for the future. This issue is especially acute for sustainable management of natural resources. It is no longer possible to manage systems so they will remain the same as in the recent past, which has traditionally been the reference point for resource managers and conservationists. We must adopt a more flexible approach to managing resources—management to sustain the functional properties of systems that are important to society under conditions where the system itself is constantly changing. Managing resources to foster resilience—to respond to and shape change in ways that both sustain and develop the same fundamental function, structure, identity, and feedbacks—seems crucial to the future of humanity and the Earth System. Resilience-based ecosystem stewardship is a fundamental shift from steady-state resource management, which attempted to reduce variability and prevent change, rather than to respond to and shape change in ways that benefit society (Table3.1). We emphasize resilience, a concept that embraces change as a basic feature of the way the world works and develops, and therefore is especially appropriate at times when changes are a prominent feature of the system. We address ecosystems that provide a suite of ecosystem services rather than a single resource such as fish or trees. We focus on stewardship, which recognizes managers as an integral component of the system that they manage. Stewardship also implies a sense of responsibility for the state of the system of which we are a part (Leopold 1949). The challenge is to anticipate change and shape it for sustainability in a manner that does not lead to loss of future options (Folke et al. 2003). Ecosystem stewardship recognizes that society's use of resources must be compatible with the capacity of ecosystems to provide services, which, in turn, is constrained by the life-support system of the planet (Fig. 1.2).

Table 3.1: Contrasts between steady state resource management, ecosystem management, and resilience based ecosystem stewardship

Steady state resource management	**Ecosystem management**	**Resilience based ecosystem Stewardship**
Reference state: historic condition	Historic condition	Trajectory of change
Manage for a single resource or species	Manage for multiple ecosystem services	Manage for fundamental social–ecological properties
Single equilibrium state whose properties can be sustained	Multiple potential states	Multiple potential
Reduce variability	Accept historical range of variability	Foster variability and diversity
Prevent natural disturbances	Accept natural disturbances	Foster disturbances that sustain social–ecological properties
People use ecosystems	People are part of the social–ecological system	People have responsibility to sustain future options
Managers define the primary use of the managed system	Multiple stakeholders work with managers to define goals	Multiple stakeholders work with managers to define goals
Maximize sustained yield and economic efficiency	Manage for multiple uses despite reduced efficiency	Maximize flexibility of future options
Management structure protects current management goals	Management goals respond to changing human values	Management responds to and shapes human values

This figure delineated above presents a framework for understanding and managing resources in a world where persistent directional changes are becoming more pronounced.

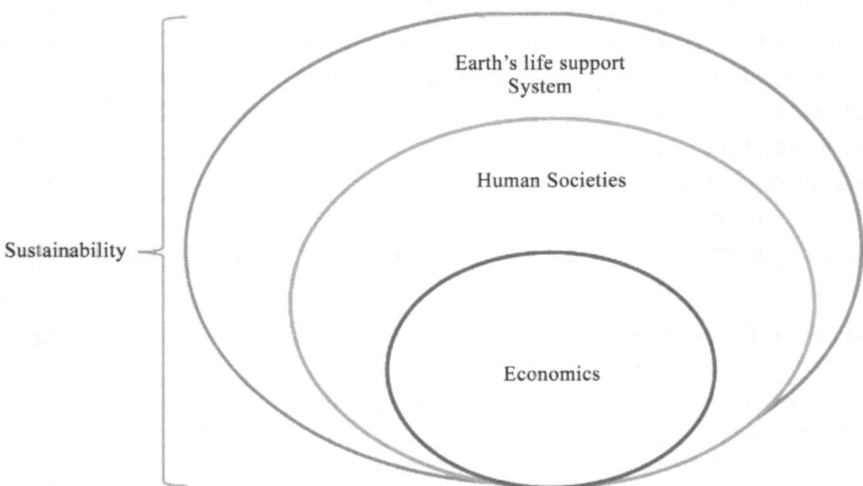

Fig. 3.2: Social–ecological sustainability requires that society's economy and other human activities not exceed the capacity of ecosystems to provide services, which, in turn, is constrained by the planet's life support system.

3.13 The 10 Elements of Agroecology

The 10 Elements of Agroecology resulted from a multi-stakeholder process intended to generate a system re-design framework to be optimized and adapted to local contexts. The 10 Elements of Agroecology framework was originally developed in 2015. Prominent themes from presentations delivered during the First International Symposium on Agroecology for Food Security and Nutrition provided an initial coherent structure: recycling, efficiency, diversity, resilience and synergies as central ecological features of agroecology. Nevertheless, calls in regional meetings for reinforcing social and political aspects of agroecology were also strong. Thus, these aspects that emerged from regional consultations were clustered under an additional five elements: co-creation of knowledge; human and social values; culture and food traditions; responsible governance; and circular and solidarity economy. Following the refinement of element names, content, and the development of a consistent storyline highlighting the interconnected nature of agroecology and its 10 Elements, the framework was finalized after several rounds of review by international and FAO experts.

The 10 Elements of Agroecology framework was launched at the Second FAO International Symposium on Agro ecology held in April 2018 and continues to evolve. In October 2018, the 10 Elements of Agroecology were supported by the FAO Committee on Agriculture (COAG) at its 26th Session as a guide to one of the ways to promote sustainable agriculture and food systems. Following the review, revision, and clearance process through FAO's governing bodies, the 10 Elements of Agroecology were approved by the 197 Members of FAO to guide FAO's vision on agroecology by the 163 session of the Council on 2-6 December 2019.

1. **Diversity:** diversification is key to agroecological transitions to ensure food security and nutrition while conserving, protecting and enhancing natural resources.
2. **Co-creation and sharing of knowledge:** agricultural innovations respond better to local challenges when they are co-created through participatory processes.
3. **Synergies:** building synergies enhances key functions across food systems, supporting production and multiple ecosystem services.
4. **Efficiency:** innovative agro-ecological practices produce more using less external resources.
5. **Recycling:** more recycling means agricultural production with lower economic and environmental costs.
6. **Resilience:** enhanced resilience of people, communities and ecosystems is key to sustainable food and agricultural systems.
7. **Human and social values:** protecting and improving rural livelihoods, equity and social well-being is essential for sustainable food and agricultural systems.
8. **Culture and food traditions:** by supporting healthy, diversified and culturally appropriate diets, agroecology contributes to food security and nutrition while maintaining the health of ecosystems.
9. **Knowledge and Governance:** sustainable food and agriculture requires responsible and effective governance mechanisms at different scales – from local to national to global.
10. **Circular and solidarity economy:** circular and solidarity economies that reconnect producers and consumers provide innovative solutions for living within our planetary boundaries while ensuring the social foundation for inclusive and sustainable development.

3.14 Ecological Resilience and Conservation Agriculture

The food choice that precipitated the COVID-19 crisis was predicted more than a decade ago,1 and the disruption induced by the virus fits a model of adaptive cycle dynamics developed nearly 50 years ago (Holling, 1973). Adaptive cycles of rapid growth, maturity, release, and reorganization are ubiquitous in social-ecological systems. Ecological resilience refers to the degree of disturbance a system can buffer before entering the collapse/release and reorganization phases. Systems can induce collapse and reorganization that affects the resilience of their subsystems as well as other systems (Sundstrom & Allen, 2019). In recent years the concept of ecological resilience has been applied most vigorously to food systems and climate change. Agriculture is affected by climate change perhaps more than any other sector of our society (National Sustainable Agriculture Coalition [NSAC], 2019). One recent policy response is the Agriculture Resilience Act (2020), which seeks to mitigate climate change by sequestering carbon and reducing other greenhouse gases through processes that enable food systems to cope with climate change by increasing soil health and, thereby, yields and profits.

The concept or metaphor of the adaptive cycle was developed by Holling (1986) to describe the different phases of system behavior in (managed) ecosystems. These systems show a tendency to repeat characteristic behavioral phases e.g. succession states. The cycle was first used to describe temperate ecosystems (e.g. boreal coniferous forests, grassland). Later it was also adapted to ecosystems in other regions of the world and even to human organizations (e.g. bureaucratic) and economics. In a traditional ecology view of the succession processes two phases are emphasized: The first is the rapid colonization of disturbed areas (exploitative phase). The second is the conservation, the slow accumulation and storage of energy and material in established (Climax) ecosystems (conservation phase). The organisms engaged in the exploitive phase are termed r-strategists the ones predominant in the conservation stage K-strategists. The r types are usually so-called pioneer species, characterized by exponential population growth – often leading to a final collapse – and excellent dispersal strategies. The K – strategists are seen to maintain their population close to some maximum sustainable population. Generally, they have less offspring (but invest more into them) and are more specialized then the r-strategists. Holling now adds two additional functions: (1) the release or "creative destruction" (a term borrowed from Schumpeter (1950)) (2) the reorganization phase. The release phase is termed O the reorganization phase a. In a two-dimensional view the adaptive cycle has two properties: The potential and the connectedness. In the K-phase of a system the accumulation (potential) of biomass and nutrients is high and tightly bound (over connected in system terms). This leads to an increasingly fragile system. Then some sudden events release the biomass and nutrients and lead to the release or O phase). Such events can be: forest fires, drought, insect pest, or intense pulses of grazing. The ensuing phase is the reorganization or a phase. In this phase the nutrient loss is minimized and the nutrient availability reorganized. The pioneer species appear and utilize accumulated and emerging energy and nutrient sources.

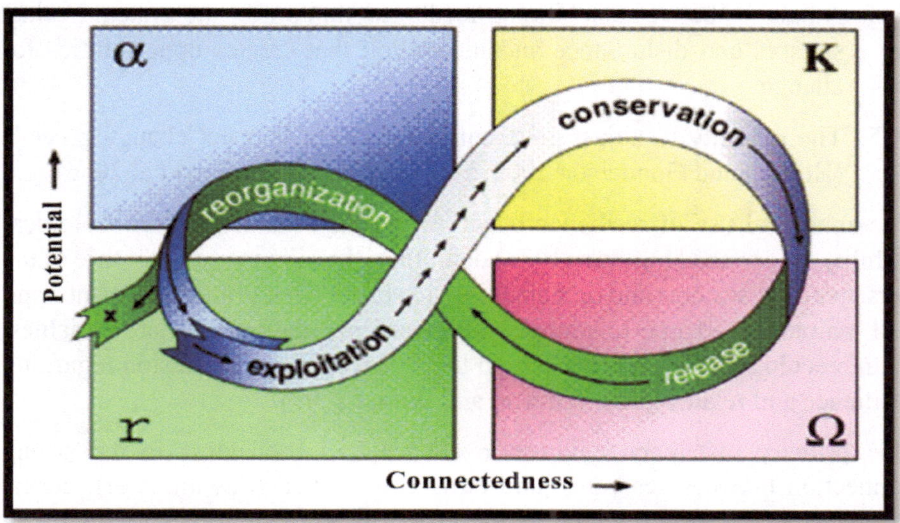

Fig. 3.3: The "lazy eight" is a stylized illustration of the four ecosystem functions (r, O, K a), and the flow of events among them. The length of the arrows indicates the speed of change in the different phases (short arrows = fast; long arrows = slow). The x-axis shows the degree of connectedness the y-axis the potential of the accumulated resources. (In Gunderson et al, 2002)

The resilience of socio-ecological systems is described by Carpenter et al (2002) as the amount of disturbance a system can absorb and still remain within the same state. The degree to which the system is capable of self-organization (versus lack of organization, or organization forced by external factors) and the degree to which the system can build and increase the capacity for learning and adaptation. The greater the resilience of the system the easier it will absorb shocks and be capable to adapt to changes. The adaptive capacity of a society is constraint by its institutions and by the environment. Social-ecological resilience is determined by the livelihood security of an individual or groups. This security involves the access and entitlement to natural resources (Berkes et al 2003). The resilience of social system is also influenced by the interaction of different levels of hierarchies.

Resilience depends on

1. Adaptive capacity
2. Biophysical and social legacies that contribute to diversity and provide proven pathways for rebuilding
3. The capacity of people to plan for the long term within the context of uncertainty and change

4. A balance between stabilizing feedbacks that buffer the system against stresses and disturbance and innovation that creates opportunities for change
5. The capacity to adjust governance structures to meet changing needs (Holling and Gunderson 2002, Folke et al.2003, Walker et al.2006)

The analytical use of resilience began in ecology, mainly within ecological stability theory and Holling's (1973) definition of resilience as a measure of the persistence of systems and of their ability to absorb change and disturbance and still maintain the same relationships between populations or state variables. Within ecology there has been a high level of conceptual confusion regarding resilience and related terms (Grimm and Wissel 1997).

One primary difference between resilience and robustness is the strong connection between resilience and other concepts such as attractors, states, steady states, stability, domains of attraction, basins of attraction, regimes, and equilibria, none of which are referenced frequently by literature on system robustness.

Conservation agriculture is mainly a concept for resource-saving agricultural crop production to achieve acceptable profits along with high and sustainable production levels subsequently conserving the environment" (FAO 2007). So, it has evolved as an alternative approach to conventional agricultural systems in which tillage is practised to varying levels to control weeds, pests, nutrient incorporation and to check soil compaction problems where conservation agriculture helps to maintain, to the extent possible, a year-round soil cover by residue retention from previous crops or incorporation from another field/crop and/or a cover crop to improve the soil quality and minimal soil disturbance by no or reduced tillage practices and through crop diversification and/or crop rotations to restrict the pest problems (Dillaha et al.).

Enhanced resilience of people, communities and ecosystems is key to sustainable food and agricultural systems. Diversified agroecological systems are more resilient – they have a greater capacity to recover from disturbances including extreme weather events such as drought, floods or hurricanes, and to resist pest and disease attack. Following Hurricane Mitch in Central America in 1998, biodiverse farms including agroforestry, contour farming and cover cropping retained 20–40 percent more topsoil, suffered less erosion and experienced lower economic losses than neighbouring farms practicing conventional monocultures.

By maintaining a functional balance, agroecological systems are better able to resist pest and disease attack. Agroecological practices recover the biological complexity of agricultural systems and promote the necessary community of interacting organisms to self-regulate pest outbreaks. On a landscape scale, diversified agricultural landscapes have a greater potential to contribute to pest and disease control functions.

Agroecological approaches can equally enhance socio-economic resilience. Through diversification and integration, producers reduce their vulnerability should a single crop, livestock species or other commodity fail. By reducing dependence on external inputs, agroecology can reduce producers' vulnerability to economic risk. Enhancing ecological and socio-economic resilience go hand-in-hand – after all, humans are an integral part of ecosystems.

3.15 Ecosystem Services and Conservation Agriculture

Conservation agriculture (CA) has emerged as a promising technology for efficient rational use of available resources and sustained productivity in the long run. By saving inputs, reducing energy usage and greenhouse gases emissions, CA-based management practices are quite viable for bringing sustenance in agricultural crop production. The CA system can provide multiple ecosystem services such as provisioning, regulating and supporting services. The regulating services include improving carbon status, and physical, chemical and biological properties of soil, which further lead to provisioning services in terms of sustained crop and water productivity. Increased soil carbon sequestration improves supporting services, namely, soil aggregation that increases available soil moisture and can be helpful for better plant growth and development. It also improves soil biodiversity both above- and below-ground. Here we focus on the potential ecosystem service benefits accrued from CA. Conservation agriculture in the long run can be a strategy for sustainable crop intensification and a climate resilient crop management system.

CA is described by three interlinked principles, along with other good agricultural practices, namely: (i) continuous no or minimal mechanical soil disturbance (implemented by the practice of no-till seeding or broadcasting of crop seeds, and direct placing of planting material into untilled soil; and causing minimum soil disturbance from any cultural operation, harvest operation or farm traffic); (ii) maintenance of a permanent biomass soil mulch cover on the ground surface (implemented by retaining crop biomass, root stocks and stubbles and cover crops and other sources of ex situ biomass); and (iii) diversification of crop species; implemented by adopting a cropping system with crops in rotations, and/or sequences and/or associations involving annuals and perennial crops, including a balanced mix of legume and non-legume crops (Kassam et al. 2018). Controlled traffic that lessens soil compaction could be another principle being pondered upon in the recent years. CA is a promising technology for efficient rational use of available resources and sustained productivity in the long-run (Das et al. 2014, Nath et al. 2017, Oyeogbe et al. 2018).

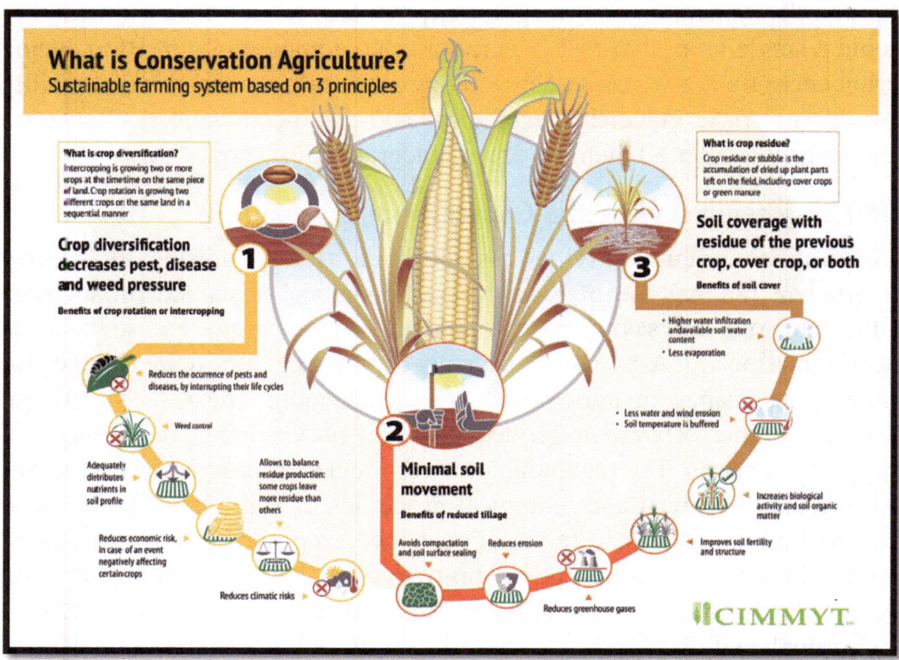

Fig. 3.4: Conservation Agriculture and its principles

There are many potential ecosystem services associated with the benefits of conservation agriculture. Ecosystem services are natural processes through which the environment produces natural resources that humans and other living species require for life (Dillaha et al. 2010). These include the benefits that human derive directly or indirectly from ecosystem functions (Costanza et al. 1997). Millennium Ecosystem Assessment (2005) has grouped ecosystem services into four broad categories: provisioning, supporting, regulating and cultural services with specific services mentioned in each category. These are presented in Fig 3.4 with some modifications/additions of specific services. Although CA was originally introduced to regulate wind and water erosion (Baveye et al. 2011), is now considered to deliver multiple ecosystem services (Oyeogbe et al. 2017, 2018).

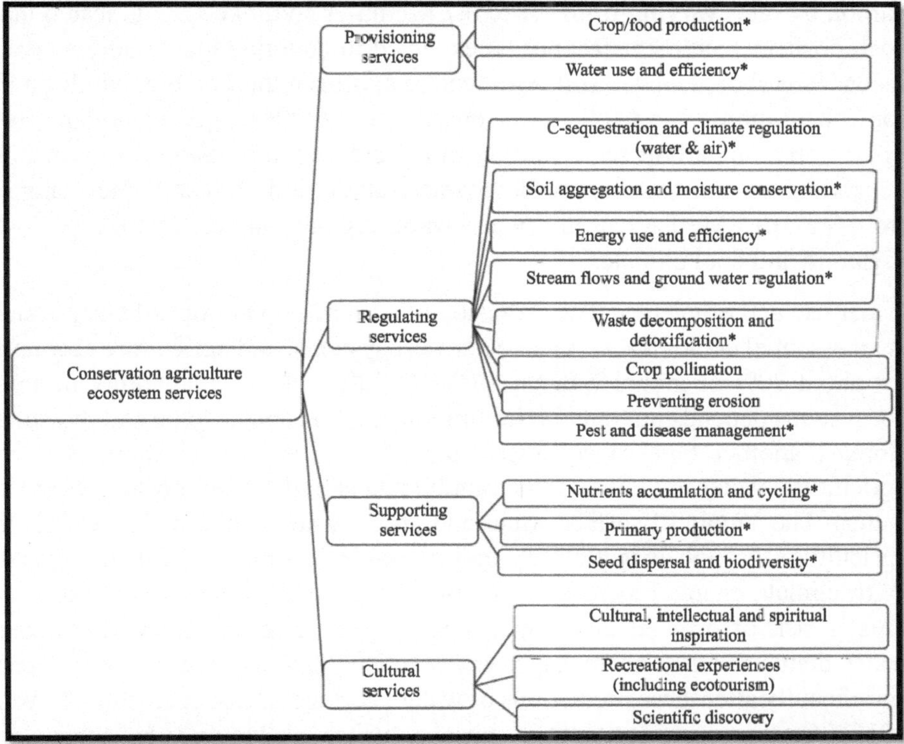

Fig. 3.5: Different ecosystem services provided by conservation agriculture (*services are positively influenced by CA) (*Source*: Millennium Ecosystem Assessment 2005, Dillaha et al. 2010, modified by Authors)

Carbon sequestration and climate regulation: One of the most positive aspects of conservation agriculture is its contribution to increase soil carbon compared with tillage-based conventional agriculture systems. Lal (2004) in his studies of the benefits of increasing soil carbon on its potential for carbon sequestration for climate regulation found that the carbon sink capacity of the world's agricultural and degraded soils is 50 percent to 66 percent of the global historic carbon loss of 42 to 78 gigatons of carbon and that improved land management practices on the world's agricultural and degraded soils could sequester 50 percent to 66 percent of the historic soil carbon loss. This is equivalent to 0.4 to 1.2 Gt C/year, or 5 percent to 15 percent of global carbon emissions. Lal also stated that the rate of increase in the Soil Organic Carbon (SOC) stock, through recommended management practices, follows a sigmoid curve, attains the maximum 5 to 20 years after adoption of recommended management practices, and continues until SOC attains a final equilibrium. Hillel and Rosenzweig (2009) also studied that the conversion to no-till farming increased soil organic carbon by rates varying from 0.1 to 0.7 Mg/ha-yr and, like Lal, indicated that such positive increments cannot be expected to continue indefinitely as well managed carbon depleted soils will tend to approach their natural equilibrium (or C saturation) state within a few decades. Lal (2004) reported similar rates of soil organic carbon sequestration in agricultural and restored ecosystems depending on soil texture, profile characteristics, and climate, which ranged from 0 to 0.15 Mg/ha-year in dry and warm regions, and 0.1 to 1.0 Mg/ha-yr in humid and cool climates.

Water use and efficiency: The provision of sufficient quantities of clean water is an essential ecological service to agro-ecosystems, and agriculture accounts for about 70% of global water use (WHO 2003). Water availability in agro ecosystems depends not only on infiltration and flow, but also on soil moisture storage, another type of ecosystem service (Power 2010). About 80% of agricultural water use comes from rainfall and stored as soil moisture (~green water). The supply of surface water and groundwater (~blue water) inputs to agriculture through irrigation are indispensable in some parts of the world. With climate change, increased variability of rainfall is predicted to lead to greater risk of drought and flood, while higher temperatures will increase water demand. On-farm management practices that target green water, can significantly alter these predictions of water shortages (Rost et al. 2009, Power 2010). Conservation agriculture improves water productivity by enhancing infiltration, reducing soil evaporation and increasing soil water holding capacity for subsequent stomatal transpiration (Dillaha et al. 2010).

Energy use and efficiency: Reduction of energy requirement is possible by reducing number of tillage operations by adoption of conservation tillage. According to Kour et al. (2011), conservation tillage could offset as much as 16% of world-wide fossil fuel emissions and can slow or prevent the loss of organic C in soil. On an average, by adopting ZT for land preparation and crop establishment in rice-wheat system of the IGP, farmers could save 36 l diesel/ha (Erenstein and Laxmi 2008). Saad et al. (2016) studied the energy auditing in CA-based maize-wheat-mungbean system and found that ZT bed planting with wheat and maize residue retention could be a substitute of the conventional agricultural system for adoption in maize-wheat-greengram cropping system in the irrigated north western Indo-Gangetic Plains. Kumar et al. (2013) showed that ZT improved the operational field capacity by 81%, specific energy by 17% and the energy usage efficiency by 13% as compared to CT.

Preventing erosion: Conservation agriculture and other forms of conservation tillage highly reduce soil erosion which has been documented in most of the studies regarding conservation agriculture. A study conducted in Mississippi (Dabney et al., 2004), it has found that, on a silt loam soil comparing no-till and chisel/disc till under corn cultivation, no-till decreased soil erosion significantly while the land was in no-till for 5 to 10 years and for the first year after no-till ended, but a year no-till was abandoned and tillage was reinstituted, the protective effects of the previous no-till were no longer significant.

Bio-diversity: The intensification of agriculture and the traditional techniques based on extensive tillage have negative impact on soils, causing degradation in the physical and chemical parameters and causing loss of biodiversity. Conservation agriculture and the related management practices have demonstrated in the last decades to be an efficient tool to combine food productivity with environmental protection around the world. So, the reduction or no-tillage, the permanent soil cover, and the crop rotation or diversification practices have discernible positive effects on soils, including the improvement of physical and chemical properties, the decrease of water run-off and wind erosion, and an increase of water retention. The use of cover crop and the organic residues on the surface enhance the stability of soils and regulate temperatures. Therefore, the biodiversity increases (Conti 2015).

3.16 Soil Conservation and Conservation Agriculture

There is a lack of knowledge about the effectiveness and efficiency of soil conservation policies in agriculture and there is a little understanding of how policy measures should be designed to encourage farmers to adopt soil conservation practices. About one-sixth of the world's land area, that is, about one-third of the land used for agriculture, has been affected by soil degradation in the historic past. While most of this damage was caused by water and wind erosion, other forms of soil degradation are induced by biological, chemical, and physical processes. Since the 1950s, pressure on agricultural land has increased considerably owing to population growth and agricultural modernization. Small-scale farming is the largest occupation in the world, involving over 2.5 billion people, over 70% of whom live below the poverty line. Soil erosion, along with other environmental threats, particularly affects these farmers by diminishing yields that are primarily used for subsistence.

Global agricultural production basically consists of food for people, feed for livestock, fibre for industry, and fuel for energy. In fact, 95% of the agricultural output is produced on cultivation and grazing lands (Hurni et al., 1996), the rest being the products of marine ecosystems. About 11% of the surface of the world's terrestrial ecosystem is used for cultivation, another 25% is used for grazing or grass-cutting, a further 28% is covered by natural and planted forests of different qualities, and about 36% is desert land (FAOSTAT, 2005).

The world's human population has increased by a factor of 2.49 since 1950 (Worldwatch Institute, 2005), causing demand for the above-mentioned products to grow even faster due to increased dietary as well as other per capita demands. Agricultural production, on the positive side, increased even faster, for example, by a factor of 2.75 between 1950 and 1985 for grain cereals (FAOSTAT, 2005), and has remained at about this level. This early increase was due to a number of factors, including advances in agricultural research and technology, plant breeding, increased inputs in minerals and fertilizer, a slight expansion of cultivated land by a factor of 1.13 (FAOSTAT, 2005), and intensification on currently cultivated land.

At present, it appears that both the area of agricultural land and the productivity potentials have reached their upper limits, with little scope to (a) further expand agricultural land sustainably and (b) develop plants that are capable of producing even more per hectare without negatively affecting people and ecology. This, however, is contested by some scientists who claim that there are still vast areas of underutilized arable land, and others who believe that biotechnology will find additional possibilities to enhance plant and animal productivity.

In the last 50 years, pressure on soil resources has been considerably accelerated in many places, along with agricultural intensification and particularly expansion. In response, there have been efforts to conserve soil to find means of agricultural production that minimize negative impacts. Much of this reproductive activity, however, is being challenged by persistent poverty, while current actions to reduce poverty in many countries often put even greater strain on agricultural land resources.

As a response, it is recognized that Conservation Agriculture (CA) may strengthen a large set of soil ecosystem services (Palm et al., 2014). CA is based on three technical principles which are (i) minimum soil disturbance, (ii) retention of crop and cover crop residues and (iii) use of diversified cropping patterns (FAO, 2014; Séguy et al., 2006). CA was promoted for providing multiple beneficial effects on soil physical (Indoria et al., 2017; Patra et al., 2019), biological (Lienhard et al., 2013; Mathew et al., 2012; Souza et al., 2018) and chemical properties (Ivy et al., 2017; Parihar et al., 2018; Ranaivoson et al., 2017). Amongst soil chemical properties, soil organic carbon (SOC) was widely studied under CA, due to its potential to adapt farming systems to climate change and variability, and mitigate the effects of climate change through carbon sequestration (Lal, 2013). Globally, these studies allowed to provide process-based analysis of the soil physical, biological and/or chemical compartments, as affected by CA practices. However, such analyses do not seem sufficient to propose consistent soil health assessments.

Soil health was defined by Kibblewhite et al. (2008) as the capacity of a soil to produce a good quantity and quality food and fibre together with the delivery of other ecosystem services. However, in most studies in the literature, all variables used to assess soil health are treated without proper consideration of their role along the soil health causal chain. Indeed, there is often no consideration of the cause/consequence relation between soil properties, soil functions and soil ecosystem services (Ivy et al., 2017; Lienhard et al., 2013; Thierfelder and Wall, 2012). while the realization of soil properties leads to the expression of soil functions which allows a soil to provide ecosystem services (Kibblewhite et al., 2008; Su et al., 2018). The lack of consensus on this causal chain, the different interaction between soil properties and the addition of separated soil health indicators lead to different considerations of the components of soil systems which may lead to misleading conclusions on soil functioning assessment. To fill this gap, integrative approaches based on the results of soil physical-biological-chemical interactions should be implemented (Kibblewhite et al., 2008; Vogel et al., 2018; Rinot et al., 2019). Finally, assessments should be easily transferable to smallholder farmers, development practitioners and land managers who are the first beneficiaries of soil health (Idowu et al., 2008), but most studies rely on laboratory tools, protocols, and complicated data-analysis interpretation which may limit the further dissemination of new practices (Bünemann et al., 2018).

In semi-arid and arid regions, variable pattern of rainfal and scarcity are significant constraints to agricultural productivity and here, conservation agriculture would be expected to increase infiltration, soil moisture storage and utilization, and also water use efficiency. Unfortunately, in many low rainfall regions with significant dry seasons, it is difficult to maintain an effective soil organic matter cover because of competing uses as livestock feed or sometimes as fuel. In these cases, conservation agriculture may not work because of removal of the organic cover, even with no till, results in bare soil and the formation of a soil crust, which will decrease the potency for water infiltration. This was stated by Rockström et al. (2009) in case of conservation farming strategies for arid and semi-arid agriculture in East and Southern Africa. It was seen that yields were higher with conservation agriculture than conventional tillage during drier rainy seasons but there was not much difference in yields during wetter rainy seasons suggesting that conservation farming in savannah agro-ecosystems may foremost function as a water harvesting system, which will enhance the ability of crops to check dry spells.

Conservation agriculture avoid mechanical soil disturbance to the extent possible. Avoid soil compaction beyond the elasticity of the soil. Maintain or improve soil organic matter during rotations until reaching an equilibrium level. Maintain organic cover through crop residues and cover crops to minimize erosion loss by wind and/or water.

3.17 Water Conservation and Conservation Agriculture

Water and soil nutrients remain the most limiting resources for agriculture. However, rainfed and irrigated agriculture is often hindered by the depletion of soil nutrients through surface runoff and leaching during rainy phases, as well as by the scarcity of water during dry phases. Surface runoff adversely affects the availability of water, soil nutrients, and soil organic matter for plant growth and development. Percolated water (leaching) can also affect water and nutrient availability to some extent. Agricultural activities exacerbate the removal of nutrients by either of the processes. Nutrients that are removed by surface runoff are permanently lost before reaching the root zone of the plant, while nutrients that are leached below the root zone are at least temporarily lost from the root system. Concurrently, the nutrient and water components of surface runoff and percolation may also deteriorate the water quality of wells, reservoirs, and lakes. Thus, the nutrients often removed by water movement and dynamics contribute not only to water quality deterioration, but this also imply an economic loss of soil fertility to the farmer. The current approach of agricultural systems, which promotes the use of more chemical fertilizers, particularly for vegetable production, shows a wider expansion of the above risks and is becoming a serious threat to our environment.

Hence, there is a need for a paradigm shift to improve smallholder agriculture systems that promote sustainable intensification, which encourages an increase in crop productivity with minimum inputs and protects the environment at the same time. Smallholder vegetable production may also be optimized by applying conservation agriculture (CA) practices that would improve productivity with minimum inorganic inputs and minimize adverse effects on the environment. The CA system (minimum soil disturbance, complete soil cover, and proper crop rotation) has been used to improve irrigation water use efficiency and crop productivity while controlling soil nutrient losses caused by various factors. No-tillage, despite the challenges of implementation and adoption, reduces runoff, increases percolation, and enhances water holding capacity of soils, when combined with grass mulch cover and proper crop rotation. The biological decomposition of grass mulch has improved the soil quality (adding soil nutrients) and soil structure, while a no-till practice combined with a complete soil cover reduces the soil compaction in the long term, particularly in drier regions or dry phases.

The yield of cereal crops has been increased in CA systems (due to higher water infiltration) under rainfed phases of production. Water use efficiency and the yield of vegetables significantly increase under CA in dry irrigation phases of production compared with the conventional practices. Application of CA in vegetable production systems with local varieties based on market demand has greater opportunities for adoption. in supplementary irrigation vegetable production systems, CA can provide opportunities to optimize water use by decreasing irrigation water requirements and optimize nutrient use by decreasing nutrient losses through runoff and leaching. These best management practices will not only improve yield and efficiency of inputs (water and nutrient) but also decrease pollution and protect our environment.

3.18 Biodiversity Conservation and Conservation Agriculture

Unregulated urbanization, industrialization and increased human activities have led to rapid deterioration of the environment. This has severely affected the life supporting system.

Last decade or so has seen a serious concern from people of all walks of life towards various environmental issues which include global warming, biodiversity loss, depletion of ozone layer, acid rain, dumping of hazardous wastes etc. World is a beautiful place to live in because of the variety of living organisms including plants, animals and micro-organisms with whom we share this planet. The remarkable diversity of living organisms form an inseparable and significant parts of our planet. However, the ever increasing human population and the means and methods of survival that have emerged are posing serious threats to bio-diversity. Biodiversity is a natural wealth essential for human survival as much as for the survival of other species.

In addition, biodiversity in land ecosystems generally decreases with increasing altitude. The other factors that influence biodiversity are amount of rainfall and nutrient level in soil. In marine ecosystems, species richness tends to be much higher in continental shelves. India is a country of vast diversity and it is among the 12 “mega-diversity” countries in the world. Both flora and fauna, all over the world are under an assault from a variety of indiscriminate human activities. These activities are often related to rapid growth of human population, deforestation, urbanization and industrialization.

Conservation is the planned management of natural resources, to retain the balance in nature and retain the diversity. It also includes wise use of natural resources in such a way that the needs of present generation are met and at the same time leaving enough for the future generations. Conservation of biodiversity is important to: – prevent the loss of genetic diversity of a species, save a species from becoming extinct, and protect ecosystems damage and degradation.

Conservation efforts or strategies can be grouped into the following two categories: In-situ (on-site) conservation includes the protection of plants and animals within their natural habitats or in protected areas. Protected areas are land or sea dedicated to protect and maintain biodiversity. It includes National parks & sanctuaries, Biosphere reserves, sacrization of forests and lakes. Ex-situ (off-site) conservation of plants and animals outside their natural habitats. To complement in-situ conservation efforts. Ex-situ conservation is being undertaken through setting up botanic gardens, zoos, medicinal plant parks, etc. by various agencies. The Indian Botanical Garden in Howrah (West Bengal) is over 200 years old. These include botanical gardens, zoo, and gene banks, seek bank, tissue culture and cryopreservation ("freeze preservation"). Conservation of biodiversity is like an insurance policy for future as it will ensure continuity of food chains, sustainable utilization of life support systems on earth.

3.19 Barriers to Conservation Agriculture

Conservation agriculture (CA) poses a challenge both for the scientific community and farmers to overcome the traditional mindset. Worldwide CA has spread mostly in rainfed farming system, but in India largely restricted to irrigated agriculture. The main reasons are

(i) Heterogeneous soils, high slopes and physical constraints like crusting and hard pans that reduce seed germination, decrease emergence and causes poor crop stand.

(ii) Non ¬availability of crop residues as it is used as fodder.

(iii) Need for efficient in ¬situ rain water harvesting through permanent bed and furrow system.

(iv) Weed management between two crop.

A mental change of farmers, technicians, extension personnel and researchers away from soil degrading tillage operations towards sustainable production systems like no tillage is necessary to obtain changes in attitudes of farmers (Derpsch, 2001). Hobbs and Govaerts (2010) however, noted that probably the most important factor in the adoption of CA is overcoming the bias or mindset about tillage. It is argued that convincing the farmers that successful cultivation is possible even with reduced tillage or without tillage is a major hurdle in promoting CA on a large scale. In many cases, it may be difficult to convince the farmers of potential benefits of CA beyond its potential to reduce production costs, mainly by tillage reductions. CA is now, considered a route to sustainable agriculture. Spread of conservation agriculture, therefore, will call for scientific research linked with development efforts. The following are a few important constraints which impede broad scale adoption of CA.

- **Lack of appropriate seeders especially for small and medium scale farmers:** Although significant efforts have been made in developing and promoting machinery for seeding wheat in no till systems, successful adoption will call for accelerated effort in developing, standardizing and promoting quality machinery aimed at a range of crop and cropping sequences. These would include the development of permanent bed and furrow planting systems and harvest operations to manage crop residues.
- **The wide spread use of crop residues for livestock feed and fuel:** Specially under rainfed situations, farmers face a scarcity of crop residues due to less biomass production of different crops. There is competition between CA practice and livestock feeding for crop residue. This is a major constraint for promotion of CA under rainfed situations.
- **Burning of crop residues:** For timely sowing of the next crop and without machinery for sowing under CA systems, farmers prefer to sow the crop in time by burning the residue. This has become a common feature in the rice-wheat system in north India. This creates environmental problems for the region.
- **Lack of knowledge about the potential of CA to agriculture leaders, extension agents and farmers:** This implies that the whole range of practices in conservation agriculture, including planting and harvesting, water and nutrient management, diseases and pest control etc. need to be evolved, evaluated and matched in the context of new systems.
- **Skilled and scientific manpower:** Managing conservation agriculture systems, will call for enhanced capacity of scientists to address problems from a systems perspective and to be able to work in close partnerships with farmers and other stakeholders. Strengthened knowledge and information sharing mechanisms are needed.

3.20 Challenges in Conservation Agriculture

Conservation agriculture as an upcoming paradigm for raising crops will require an innovative system perspective to deal with diverse, flexible and context specific needs of technologies and their management. Conservation agriculture R&D (Research and Development), thus will call for several innovative features to address the challenge. Some of these are:

(a) **Understanding the system** – Conservation agriculture systems are much more complex than conventional systems. Site specific knowledge has been the main limitation to the spread of CA system (Derpsch, 2001). Managing these systems efficiently will be highly demanding in terms of understanding of basic processes and component interactions, which determine the whole system performance. For example, surface maintained crop residues act as mulch and therefore reduce soil water losses through evaporation and maintain a moderate soil temperature regime (Gupta and Jat, 2010). However, at the same time crop residues offer an easily decomposable source of organic matter and could harbour undesirable pest populations or alter the system ecology in some other way. No-tillage systems will influence depth of penetration and distribution of the root system which, in turn, will influence water and nutrient uptake and mineral cycling. Thus the need is to recognize conservation agriculture as a system and develop management strategies.

(b) **Building a system and farming system perspective** – A system perspective is built working in partnership with farmers. A core group of scientists, farmers, extension workers and other stakeholders working in partnership mode will therefore be critical in developing and promoting new technologies. This is somewhat different than in conventional agricultural R&D, the system is to set research priorities and allocate resources within a framework, and little attention is given to build relationships and seek linkages with partners working in complementary fields.

(c) **Technological challenges** – While the basic principles which form the foundation of conservation agriculture practices, that is, no tillage and surface managed crop residues are well understood, adoption of these practices under varying farming situations is the key challenge. These challenges relate to development, standardization and adoption of farm machinery for seeding with minimum soil disturbance, developing crop harvesting and management systems.

(d) **Site specificity** – Adapting strategies for conservation agriculture systems will be highly site specific, yet learning across the sites will be a powerful way in understanding why certain technologies or practices are effective in a set of situations and not effective in another set. This learning process will accelerate building a knowledge base for sustainable resource management.

(e) **Long-term research perspective** – Conservation agriculture practices, e.g. no-tillage and surface maintained crop residues result in resource improvement only gradually, and benefits come about only with time. Indeed, in many situations, benefits in terms of yield increase may not come in the early years of evaluating the impact of conservation agriculture practices. Understanding the dynamics of changes and interactions among physical, chemical and biological processes is basic to developing improved soil-water and nutrient management strategies (Abrol and Sangar, 2006). Therefore, research in conservation agriculture must have longer term perspectives.

3.21 Awareness, Knowledge and Action for the Restoration of Ecosystems

The world is facing severe challenges. Billions of people around the world are suffering the consequences of the climate emergency, food and water insecurity, and the COVID-19 pandemic. Ecosystems are an indispensable ally as we meet these challenges. Protecting them and managing their resources in a sustainable manner is essential. But just increasing the protection and sustainable management of our remaining natural landscapes and oceans will not be enough: the planet's degraded ecosystems and the huge benefits that they provide must also be restored.

By declaring the UN Decade on Ecosystem Restoration, governments have recognized the need to prevent, halt and reverse the degradation of ecosystems worldwide for the benefit of both people and nature. The 2021–2030 timeline underlines the urgency of the task. Without a powerful 10-year drive for restoration, we can neither achieve the climate targets of the Paris Agreement, nor the Sustainable Development Goals.

The UN Decade on Ecosystem Restoration provides a unique opportunity to transform food, fibre and feed production systems to meet the needs of the 21st century, and to eradicate poverty, hunger and malnutrition. This we seek to achieve through effective and innovative landscapes and seascapes management that prevents and halts degradation, and restores degraded ecosystems. The restoration of forest landscapes, farming, livestock and fish-producing ecosystems should primarily contribute to restoring them to a healthy and stable state, so that they are able to provide ecosystems services and support human needs for sustainable production and livelihoods.

For example, around one third of the world's farmland is degraded, about 87 per cent of inland wetlands worldwide have disappeared since 1700, and one third of commercial fish species are overexploited. Degradation is already affecting the well-being of an estimated 3.2 billion people – that is 40 per cent of the world's population. Every single year, we lose ecosystem services worth more than 10 per cent of our global economic output.

If we can manage to reverse this trend, massive gains await us. Reviving ecosystems and other natural solutions could contribute over one third of the total climate mitigation needed by 2030. Restoration can also curb the risk of mass species extinctions and future pandemics. Agroforestry alone could increase food security for 1.3 billion people.

Restoration on a global scale requires sustained investments. But there is growing evidence that it more than pays for itself. For example, restoring coral reefs to good health by 2030 could yield an extra USD 2.5 billion a year for both Mesoamerica and Indonesia; having doubled its forest cover since the 1980s, Costa Rica has seen ecotourism grow to account for 6 per cent of GDP.

While restoration science is a youthful discipline, we already have the knowledge and tools we need to halt degradation and restore ecosystems. Farmers, for instance, can draw on proven restorative practices such as sustainable farming (conservation agriculture) and agroforestry. Landscape approaches that give all stakeholders – including women and minorities – a say in decision-making are simultaneously supporting social and economic development and ecosystem health. And policy makers and financial institutions are realizing the huge need and potential for green investment.

Ecosystem restoration is the process of halting and reversing degradation, resulting in improved ecosystem services and recovered biodiversity. Ecosystem restoration encompasses a wide continuum of practices, depending on local conditions and societal choice. Depending on objectives, restored ecosystems can follow different trajectories:

- from degraded natural to more intact natural ecosystems (often by assisting natural regeneration)
- from degraded, modified ecosystems to more functional modified ecosystems (e.g. restoration of urban areas and farmlands)
- from modified ecosystems towards more natural ecosystems, providing that the rights and needs of people who depend on that ecosystem are not compromised.

Approaches, Principles and Tools for Restoration

Ecosystem restoration encompasses a wide variety of approaches that contribute to conserving and repairing damaged ecosystems (UNEP and FAO 2020). This may involve active restoration or the removal of drivers of degradation to 'passively' promote natural regeneration. Whatever the approach, restoration requires time, resources, knowledge, enabling policies and governance if it is to contribute to human well-being, economic development, climate stability and biodiversity conservation.

Many restoration projects and programmes have underperformed in the past. Adhering to standardized principles and guidelines, as well as continued monitoring and an adaptive approach, can ensure better performance and impact.

Restoration Approaches

- ***Ecological Restoration:*** Process- Assisting the recovery of a terrestrial, freshwater or marine ecosystem that has been degraded, damaged, or destroyed.

  Intended end point- Transition from degraded ecosystem to a reference ecosystem, which may be a natural or a cultural one.

  (Source: Gann et al. 2019)

- ***Forest and Landscape Restoration:*** Process- Reversing the degradation of soils, agricultural areas, forests and watersheds thereby regaining their ecological functionality.

  Intended end point- Restoring multiple ecological, social and economic functions across a landscape and generating a range of ecosystem goods and services that benefit multiple stakeholder groups.

  (Source: Besseau et al. 2018)

- ***Restoration of Aquatic Production Ecosystems:*** Process- Maintaining ecosystem structure and function to support food provisioning, while minimizing impacts, rather than restoring ecosystems to an initial state before production activity started.

  Intended end point- Large/oceanic marine ecosystems supporting or affected by direct and indirect impacts of fishing gears and fisheries production; recovery through changes in fishing methods and gear modification to rebuild fish stocks and reduce adverse impacts on the environment. Specificities shown for both freshwater ecosystems and coastal ones with linkages to fisheriesand aquaculture.
  (Source: FAO 2020c)

- ***Regenerative Agriculture:*** Process- Farming that uses soil conservation as the entry point to regenerate and contribute to multiple provisioning, regulating and supporting services. Intended end point- Enhancing environmental, social and economic dimensions of sustainable food production. Soil carbon, soil health and on-farm biodiversity are restored.

  (Sources: Schreefel et al. 2020; Giller et al. 2021)

- ***Rewilding:*** Process- Rebuilding, following major human disturbance, a natural ecosystem by restoring natural processes and the complete or near complete food-web at all trophic levels as a self-sustaining and resilient ecosystem using biota that would have been present had the disturbance not occurred.

  Intended end point- No pre-defined end point. Functioning native ecosystems complete with fully occupied trophic levels that are natureled across a range of landscape scales.

  (Source: IUCN CEM, n.d.)

4

Research Setting

4.1 Research Setting

In any social science research, it is hardly possible to conceptualize and perceive the data and interpret the data more accurately until and unless a clear understanding of the characteristics in the area and attitude or the behavior of people is at commend of the interpreter who intends to unveil an understanding of the implication and behavioral complexes of the individual who live in the area under reference and form a representative part of the larger community. The socio-demographic background of the local people in a rural setting has been critically administered in this section. A research setting is a surrounding in which input and elements of research are contextually imbibed, interactive and mutually contributive to the system performance. Research setting is immensely important in the sense because it is characterizing and influencing the interplays of different factors and components.

The present study was taken up in the Nadia and Hooghly district under alluvial zone and Coochbehar and Alipurduar district under terai zone of West Bengal. The study was carried out in Haringhata and Chakdaha blocks from Nadia and Balagarh block from Hooghly ditrict. In terai zone, the study was conducted in Coochbehar I, Coochbehar II and Dinhata blocks from Coochbehar district and Alipur I and Falakata blocks from Alipurduar district. A brief description of the whole study area has been provided in this chapter.

4.2 Area of Study

4.3 Profile of the State, West Bengal

The state of West Bengal was formed on 26th January, 1951. Kolkata (formerly known as Calcutta), popularly known as 'the city of joy' which once served as the capital of the country in British period, still holds its glory as the capital of the state. Presently, the state has been bifurcated in 23 districts under five divisions – Burdwan division, Jalpaiguri division, Presidency division, Medinipur division and Malda division. The districts are – Alipurduar, Bankura, Birbhum, Cooch behar, Darjeeling, Dakshin Dinajpur, Hooghly,

Howrah, Jalpaiguri, Jhargram, Kolkata, Kalimpong, Malda, Murshidabad, Nadia, North 24 Parganas, Paschim Bardhaman, Paschim Medinipur, Purba Bardhaman, Purba Medinipur, Purulia, South 24 Parganas and Uttar Dinajpur. The state stretches over an area of 88,752 sq. km. with 66 sub-divisions, 121 municipalities, 6 municipal corporations, 341 development blocks, 40, 782 Mouzas, 3349 Gram Panchayats and 37, 469 inhabited villages. West Bengal is on the eastern bottleneck of the country, stretching from the Himalayas in the north to Bay of Bengal in the south. The state extends between 86^{0}00' and 89^{0}83' Eastern longitude and 21^{0}86' and 27^{0}13' North latitude. It shares its northern boundary with Sikkim and parts of Nepal and Bhutan, eastern boundary with Bangladesh, western boundary with states of Orissa, Bihar and Jharkhand and to the southern part of the state lies the Bay of Bengal. (www.wb.gov.in)

4.3.1 Demographic Features

As per the census of 2011, the total population of the state is 91, 276, 115 of which 46, 809, 027 and 44, 467, 088 are male and female respectively. It has a share of 7.54% in total country's population with a population density of 1028 per sq. km. 68. 13% of the total population of the state counts to its rural population. The sex ratio of the state is 950 per thousand. Per capita income at current price is Rs. 61,352.35 as on August, 2014. The literacy rate of the state is 76.26 per cent, which is higher than the national average. The percentages of total literate males are 81.69 per cent and that of females are 70.54 per cent. (www.wb.gov.in)

4.3.2 Climate of West Bengal

The climate of the state comes under two distinct types. The climate of northern part of the state is characterized by sub-tropical monsoon, mild to dry winter and hot summer; whereas the southern or the Gangetic part of the state experiences a tropical savanna hot and dry winter. However, northern most part viz., Darjeeling district has a climatic type of tropical upland, mild and dry winter and short warm summer. The year may be divided into four seasons. The winter season stretches from December to February, followed by the pre-monsoon season from March to May. The period from June to the middle of September constitutes the southwest monsoon and the period from the latter half of September to November is the post monsoon period. The period from December to February is generally very unpleasant due to low temperatures over the sea except in the coastal belt. In the hot weather season from March to May, weather is dry and uncomfortable in the interior. Due to lower temperatures, the hilly regions are however, comparatively less uncomfortable. Weather tends to be oppressive during June due to high humidity and temperature. The rest period of the monsoon is fairly comfortable due to reduced day temperatures, although humidity continues to be high. Day temperatures are more or less uniform over the plains during the monsoon and post monsoon seasons (increases westwards in pre-monsoon and southwards in winter). In the state, the night minimum temperatures are lower in higher latitudes. Both day and night temperatures are lower at high level stations than over the plains. May and April are the hottest months with the mean maximum temperatures of 35.6°C and 33.8°C in the Gangetic and sub-Himalayan West Bengal respectively. The relative humidity is generally high during the period from July to September. It is about 80% in June rising to about 83%-85% in July, August and September in the morning. The diurnal variation of relative humidity is least during monsoon season. The relative humidity is lowest during the summer afternoons when it becomes 6 about 38 to 50% at the plain.

4.3.3 Agriculture Scenario of the State

West Bengal is an agrarian state. More than about 72% of the rural population of the state still earns their bread from agriculture. The total cultivable area in the state is contributes to 65.65% of the total area. The net sown area of the state is 5.234mha. Although the topography of West Bengal varies hugely, blessed by river Ganga, fertile alluvium predominates over a large area making the soil of the state highly productive. Moreover due to prevailing of both tropical and temperate climatic conditions, diversification in agriculture is another important feature of the state. Seeing its high potential in the agriculture sector, the central government has included the state as one of the seven states in "Bringing Green Revolution to Eastern India (BGREI)" Programme of 2011.

Table 1: Agricultural statistics of West Bengal

1.	Total reported area	8.68 mha
2.	Total cultivable area	5.65 mha
3.	Total forest area	1.17 mha
4.	Cultivable wasteland	16940 ha
5.	Permanent pasture and other grazing land	2370 ha
6.	Gross cultivated area	9.6 mha
7.	Net cultivated area	5.2 mha

The average cropping intensity of the state is 185% which is much higher than the national average (142%). The fertilizer consumption ratio (N, P, K ratio) of the state is 2.3:1.22:1 which is much lower than the recommended dose of 4:2:1, which clearly indicates the imbalance in nutrient supply to the plants, lowering the production and productivity. This discrepancy in fertilizer consumption also shows the nutrient deficiency in the soils of West Bengal. Besides this intensive cultivation, nutrient imbalance and nutrient deficiencies, reliance of the farmers on chemical fertilizers has increased the acidity of the soil in many parts of the state. In addition to this, the southern part of the state, the *Sunderbans* region faces problems of soil salinity because of being exposed to the brackish water of the Bay of Bengal. The major crops of the state includes food crops like paddy, wheat, maize, pulses like lentil, *mung,* arhar, gram, lathyrus etc., potato, vegetables viz. brinjal, tomato, okra, cabbage, cauliflower, cucurbits, onion, etc. and other leafy vegetables and oilseeds viz. rapeseed and mustard, sesame, linseed and others. Jute is the most important cash crop of the state. The state is also the house of the world famous Darjeeling tea. Hooghly and Burdwan districts of the state are considered to be most productive region which also forms the 'Rice Bowl' for the state. The other agriculturally rich districts are Nadia, Medinipore, both 24 Paraganas, Coochbehar etc. The state is also rich in horticulture, producing mango, banana, pineapple, papaya, guava, jackfruit, litchi, citrus, sapota etc.

The state also houses different livestock like cattle, buffaloes, sheep, pig, goat, in a considerable number.

Description of Agro-ecological Zones

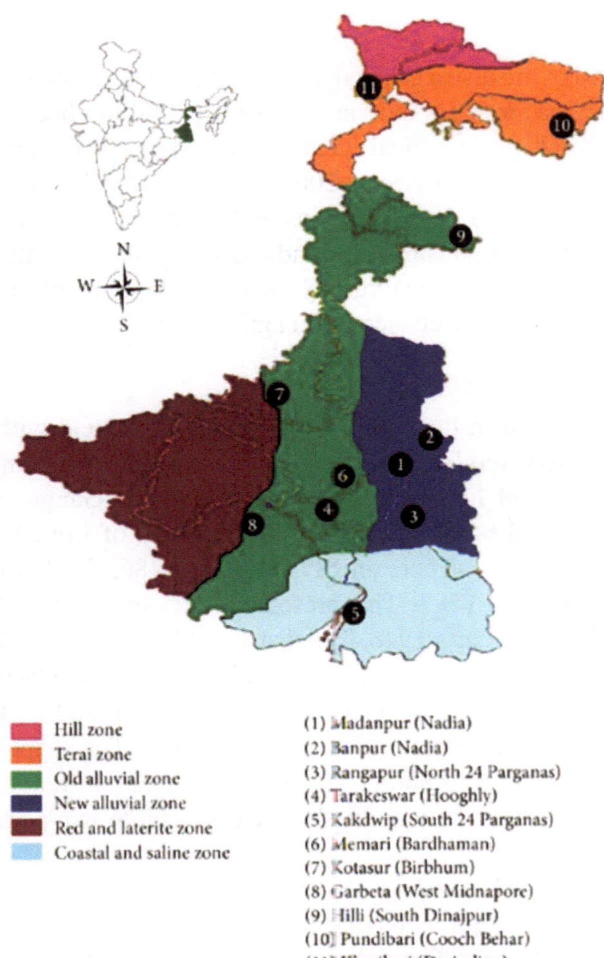

West Bengal is divided into 6 agro-ecological zones with unique and distinct features. The 6 agro ecological zones are – Hill zone, Terai zone, Old alluvial zone, New alluvial zone, Red and lateritic zone and Coastal and saline zone. The present study was conducted in New alluvial zone and Terai zone.

The New alluvial zone comprises of the districts of Nadia, Howrah, Hooghly, North 24 parganas, Kolkata, parts of Bardhaman and parts of South 24 parganas. The zones stretches over an area of 15.304 lakh ha, which is 55.7% of the total geographical area of the state. The area experiences an annual precipitation of 1350-1450mm. The average temperature ranges between 35 degree Celsius to 15 degree Celsius. The soil texture is loamy to clay in nature. The region being blessed with the fertile alluvium of the Ganges, grows a variety of agricultural crops from rice, wheat, mustard, potato, seasonal vegetables as well as cash crops like jute and mesta.

The terai region comprises of the districts of Jalpaiuri, Coochbehar, Alipurduar and Uttar Dinajpur. The zone stretches over an area of 0.17 million hectares, covering 1.9% of the total geographical area of the state. The annual precipitation in this region ranges between 2000-3200mm of rainfall and the temperature varies between 30 degree Celsius and 10 degree Celsius or below. The soil texture of this region is mainly sandy to sandy loam in nature. The region is gradually becoming agriculturally rich by adopting different sustainable agricultural practices like conservation agriculture.

Description of Nadia district

The district of **Nadia** is situated in the heart of the Bengal delta held within the arms of the Ganga, namely, the Bhagirathi on the West and Mathabhanga on the North. The entire district lies in the alluvial plain of the Ganga and its tributaries. Nadia district takes its name from the memory of Lord Shri Chaitanya Mahaprabhu who was born here on 18th February, 1486. The British district of Nadia was formed in 1787. The present district of Nadia after partition was formed on 23rd February, 1948. The district of Nadia is bounded on the north and north-west by the district of Murshidabad. On the south-east and east it is bounded by the Republic of Bangladesh. In the south and south-east the district is bounded by the district of North 24-Parganas. The shape of the district is irregular, lying North to South. The district is about 46ft. above the mean sea-level and the tropic of cancer divides the district into two parts.

Demography of Nadia District

The district has an area of 3927 sq kms having a population of 46,04,827 as per Census 2001. Out of that SC & ST population are 13,65,985 and 1,13891 respectively. The district has 17 Panchayet Samities consist of 187 Gram Panchyets and 8 Municipalities. Total number of Police Stations in the district is 19. The density of population in this district is 1173 persons per sq km. Nadia district has 950 females as against 1000 males. The majority of the people of the district speaks Bengali followed by Hindi, Santali and other. About 73.75% are Hindus and 25.42% are Muslims. In the district of Nadia the percentage of literacy by sex is 72.30 (Male) & 59.60 (Female) as per Cencus 2001. The important rivers of the district are Bhagirathi, Churni, Mathabhanga, Ichamati and jalangi.

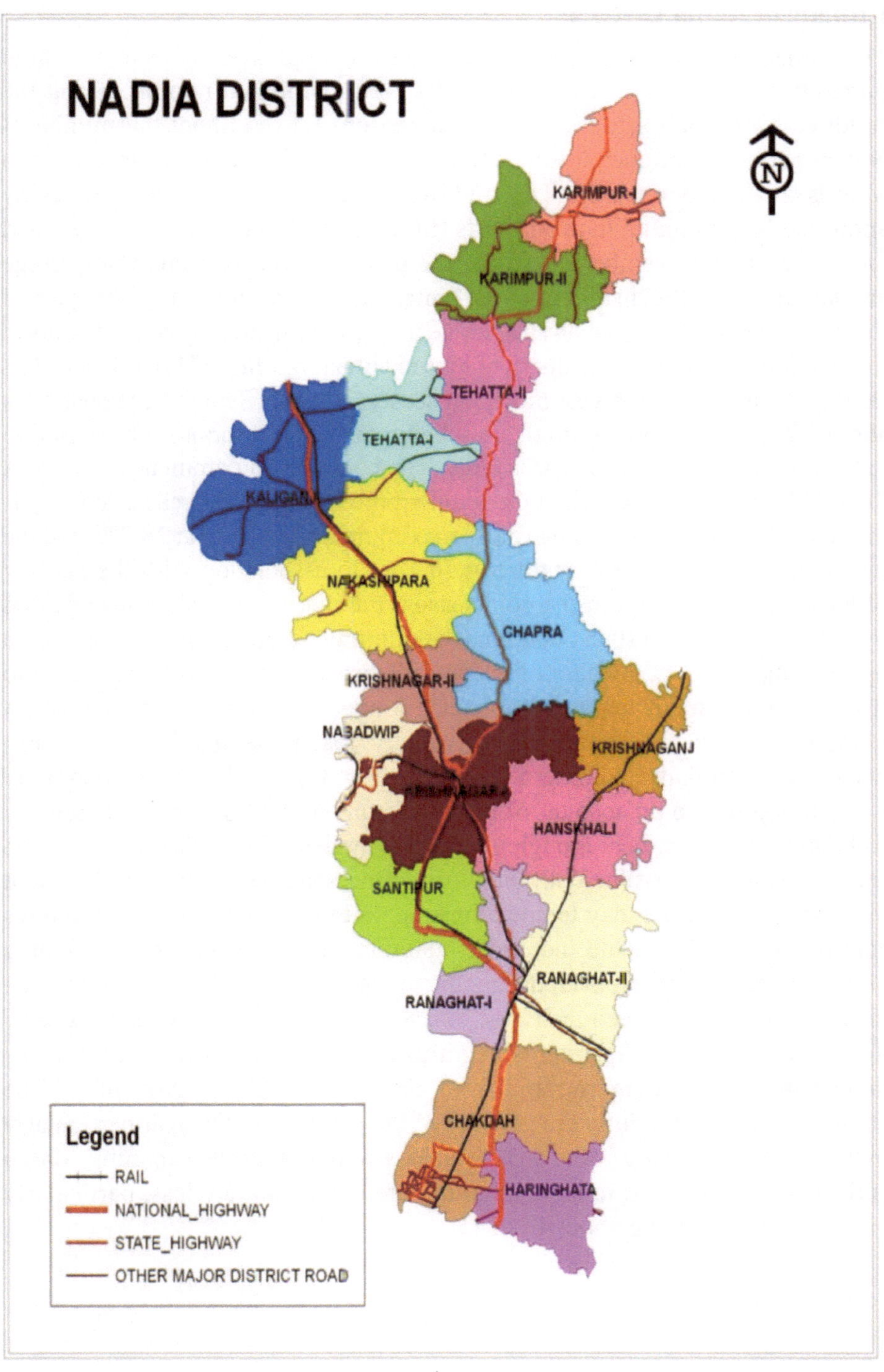
NADIA DISTRICT
N
KARIMPUR-I
KARIMPUR-II
TEHATTA-II
TEHATTA-I
KALIGANJ
NAKASHIPARA
CHAPRA
KRISHNAGAR-II
NABADWIP
KRISHNAGANJ
HANSKHALI
SANTIPUR
RANAGHAT-II
RANAGHAT-I
CHAKDAH
HARINGHATA
Legend
RAIL
NATIONAL_HIGHWAY
STATE_HIGHWAY
OTHER MAJOR DISTRICT ROAD

Climate of Nadia District

The climate of this district is characterized by an oppressive hot summer, high humidity nearly all the year round and a well distributed rainfall during the southwest monsoon season. The winter season is from about the middle of November and continues till the end of February. The period from March to May is the summer season. The southwest monsoon season commences by about the beginning of June and lasts till the end of September. October and the first half of November constitute the post-monsoon season. The average annual rainfall in the district is 1369.4 mm. Haringhata in the southern part of the district gets 1698.6 mm on an average in a year while Ranaghat further north in the middle portion of the district have an annual rainfall of 1161.0 mm. The rainfall during the southwest monsoon season, June to September constitutes about 76% of the annual rainfall, July being the rainiest month. The variation in the rainfall from year to year is large. The cold season commences by about the middle of November when the temperature begins to decrease. January is the coldest month with the mean daily maximum temperature at 28.2°C and the mean daily minimum temperature at 10.7°C. In association with the passing of western disturbances in the cold season, the district is sometimes affected by cold waves and on such occasions the minimum temperature may go down up to about 1 to 3°C. By about the end of February the temperatures begin to rise. While the night temperatures reach a maximum only in the southwest monsoon season, day temperatures usually reach the maximum in April when the mean daily maximum temperature is 38.0°C. The heat in summer is often oppressive on account of high moisture content of the air. There is a welcome relief from heat though temporarily, when thundershowers occur on some days in this season. With the onset of the southwest monsoon by about the first week of June, day temperatures begin to drop but night temperatures continue to rise. With the increased humidity in the air and the continuing high night temperatures, even during the monsoon season the weather is often uncomfortable in between the rains. The southwest monsoon withdraws early in October and the temperatures begin to drop. The drop particularly in the night temperatures is more rapid from about the middle of November. The values of relative humidity are generally high throughout the year particularly in the mornings and are about 65% to 80%. But in the summer months, March and April the values of relative humidity are comparatively less, particularly in the afternoons being about 40 to 45%.

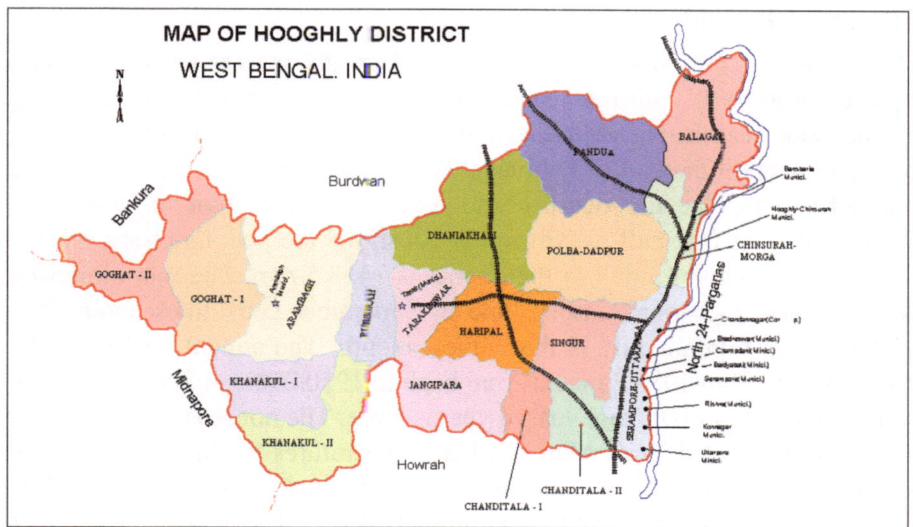

Description of Hooghly district

Hooghly district derived its name from the town of Hooghly, situated at the west bank of Hooghly River and was formed in 1795. The geographical location of the district extends between 22"39'32" – 23"01'20" northern latitude and 87"30'20" – 88"30'15" eastern longitude. To the East of the district lies the Hooghly, Bardhaman in the North, Howrah at the South, Paschim Medinipore to the West and Bankura at North-West. The maximum elevation of the district is only 200 metres. The soil type of the district is characterized by alluvial type due to well distribution of river system, which is further subdivided into categories of clayey, clayey loam and loamy. The district consists of 18 developmental blocks distributed under four sub-divisions viz. Chinsurah (5 blocks), Chandannagore (3 blocks), Sreerampore (4 blocks) and Arambagh (6 blocks). There are 210 gram panchayats in the district. Out of the total population in the district, 70% are engaged in agriculture as their primary source of income. (hooghly.nic.in)

Demography of Hooghly district

1.	Area (sq. km.)	3149
2.	Total population	58, 41, 515
3.	Rural population	68.81%
4.	Urban population	31.19%
5.	Literacy percentage	82.55
6.	Population density (per sq. km.)	1833/ sq. km.
7.	Sex ratio	958

Climate of Hooghly District

The climate of this district is characterized by an oppressive hot summer, high humidity nearly all the year round and well distributed rainfall during the monsoon season. The cold season starts by about the middle of November and continues till the end of February. The period from March to May is the summer season. The southwest monsoon season is from June to September. October and the first half of November constitute the post monsoon season. The average annual rainfall in the district is 1418.5 mm. The rainfall during the southwest monsoon season, June to September constitutes about 76% of the annual rainfall. July is the rainiest month. The annual rainfall in the district varies from 1083.2 mm at Arambagh to 1950.2 mm at Serampore. The variation in the rainfall from year to year is large. The hot season commences by about the beginning of March, when temperatures begin to rise rapidly. May is the hottest month, with the mean daily maximum temperature at 36.7°C and the mean daily minimum temperature at 24.3°C. The heat in the summer is oppressive due to high moisture content in the air. On individual days, the maximum temperature sometimes rise upto 45 to 46°C. The southwest monsoon withdraws early in October and the temperatures begin to drop. The drop in night temperatures is more rapid from about the middle of November. January is the coldest month with the mean daily maximum temperature at about 26.6°C and the mean daily minimum temperature at about 12.1°C. The values of relative humidity are generally high throughout the year in this region.

Agriculture scenario of Hooghly District

1.	Total reported area	0.31 mha
2.	Total cultivable area	0.22 mha
3.	Total forest area	530 ha
4.	Total cultivable wasteland	730 ha
5.	Permanent pasture and other grazing land	20 ha
6.	Gross cropped area	0.54 mha
7.	Net cropped area	0.21 mha
8.	Cropping intensity	256%

Description of Coochbehar District

The name *Cooch Behar* is derived from the name of the Koch or *Rajbongshi* tribes indigenous to this region. Cooch Behar is a district under the Jalpaiguri Division of the state of West Bengal. Cooch Behar is located in the northeastern part of the state and bounded by the district of Jalpaiguri and Alipurduar in the north, Dhubri and Kokrajhar district of Assam in the east and by Bangladesh in the west as well as in the south. The district forms part of the Himalayan Terai of West Bengal. The district of Cooch Behar comprises five sub-divisions:

- Cooch Behar Sadar subdivision
- Dinhata subdivision
- Mathabhanga subdivision
- Tufanganj subdivision
- Mekhliganj subdivision

Demography

According to the 2011 census Cooch Behar district has a population of 2,819,086, roughly equal to the nation of Jamaica. This gives it a ranking of 136th in India (out of a total of 739). The district has a population density of 833 inhabitants per square kilometre (2,160/sq mi). Its population growth rate over the decade 2001–2011 was 13.86%. Koch Bihar has a sex ratio of 942 females for every 1000 males, and a literacy rate of 75.49%. With 50.1% of the population, Cooch Behar is the district with the highest proportion of Scheduled Castes in the country as per the 2011 census.

Agriculture Scenario of the District

Agriculture is the primary mode of living in the district. The entire Cooch Behar district has fertile soil and around half of the cultivated land in the district is cropped twice or more. Paddy (rice) and jute are the largest producing crops, followed by potatoes, vegetables and pulses. There are 23 teagardens on glided slopes. There are some coconut, areca nut and betel leaf plantations. 77.6% of the land holdings are marginal. In 2012–13, there were 51 fertiliser depots, 2 seed stores and 64 fair price shops in the Cooch Behar ICD block. In 2012–13, the Cooch Behar I CD block produced 38,493 tonnes of Aman paddy, the main winter crop, from 19,142 hectares, 36,211 tonnes of Boro paddy (spring crop) from 9,871 hectares, 298tonnes of Aus paddy (summer crop) from 175 hectares, 8,576 tonnes of wheat from 3,413 hectares,183 tonnes of maize from 75 hectares, 164,563 tonnes of jute from 11,899 hectares, 91,495 tonnes of potatoes from 3,265 hectares and 1,255 tonnes of sugar cane from 12 hectares. It also produced pulses and oilseeds. In 2012–13, the total area irrigated in the Cooch Behar I CD block was 5,121 hectares, out of which200 hectares were irrigated by private canal water, 355 hectares by tank water, 814 hectares by river lift irrigation, 314 hectares by deep tube wells, 2,304 hectares by shallow tube wells, 97 hectares by open dug wells, 1,037 hectares by other means.

Description of Alipuduar District

Alipurduar District is the 20th district in the state of West Bengal, India. The district has its headquarters at Alipurduar. It was made a district by bifurcating Jalpaiguri district on 25 June 2014. It consists of Alipurduar municipality, Falakata municipality and six community development blocks: Madarihat–Birpara, Alipurduar–I, Alipurduar–II, Falakata, Kalchini and Kumargram. The six blocks contain 66 gram panchayats and nine census towns.

Area

Apart from the Alipurduar municipality and Falakata municipality, the district contains eight census towns and rural areas of 66 gram panchayats under six community development blocks: Madarihat–Birpara, Alipurduar–I, Alipurduar–II, Kalchini, Falakata and Kumargram. Geographically the district lies in between 26.4°N to 26.83°N and 89°E to 89.9°E.

Demographics

As of the 2011 census, Alipurduar district has a population of 1,491,250. Scheduled Castes and Scheduled Tribes make up 456,706 and 382,112 which is 30.62% and 25.62% of the population respectively.

ALIPURDUAR
C. D. BLOCK/TEHSIL MAP
BHUTAN
Jalpaiguri
Madarihat
Kalchini
Falakata
Kumargram
Alipurduar - II
Assam
BANGLA-
-DESH
Koch Bihar
Legend
International Boundary
State Boundary
District Boundary
C.D.Block/Tehsil Boundary
Copyright © 2015 www.mapsofindia.com
(Last Updated on 15th Sep 2015)

5

Research Methodology

Research methodology is a detailed plan of investigation and the blue print of procedure for carrying out the research. In this chapter, discussion on the methodology has been made to understand the concepts, methods and techniques, which are utilized to design the study, collect information, analyzing data and interpreting the findings for revelation of truth and formulation of theories. The entire chapter has been broken up under following sub-heads for easy understanding:

1. Locale of research
2. Sampling design
3. Pilot study
4. Variables and their measurements
5. Methods of data collection
6. Statistical tools used for analysis of data

5.1. Locale of Research

The present study has been conducted in two agro-ecological zones of the state West Bengal. Villages from Haringhata block and Chakdaha block in Nadia district and from Balagarh block in Hooghly district were selected from the new alluvial agro- ecological zone to conduct the study. In Terai agro ecological zone, the study was conducted in selected villages of Coochbehar I, Coochbehar II and Dinhata II blocks of Coochbehar district and in selected villages of Falakata and Alipurduar I blocks of Alipurduar district.

- The characters and the factors under study have been well discernible to this area
- The researcher's close familiarity with respect to area, people, officials and local dialects

- The ample opportunity to generate relevant data due to the close proximity of the area with the research and extension wing of the State Agricultural Universities;
- The highly cooperative and responsive respondents
- The profuse scope to get relevant information regarding perception on management of water resources, application of nutrients, improving soil organic carbon and problems and prospects of practicing conservation agricultural practices
- Experienced, well versed, venturesome, enthusiast and risk bearing farmers
- Scope of establishing a sustainable and profitable farming opportunities in the area

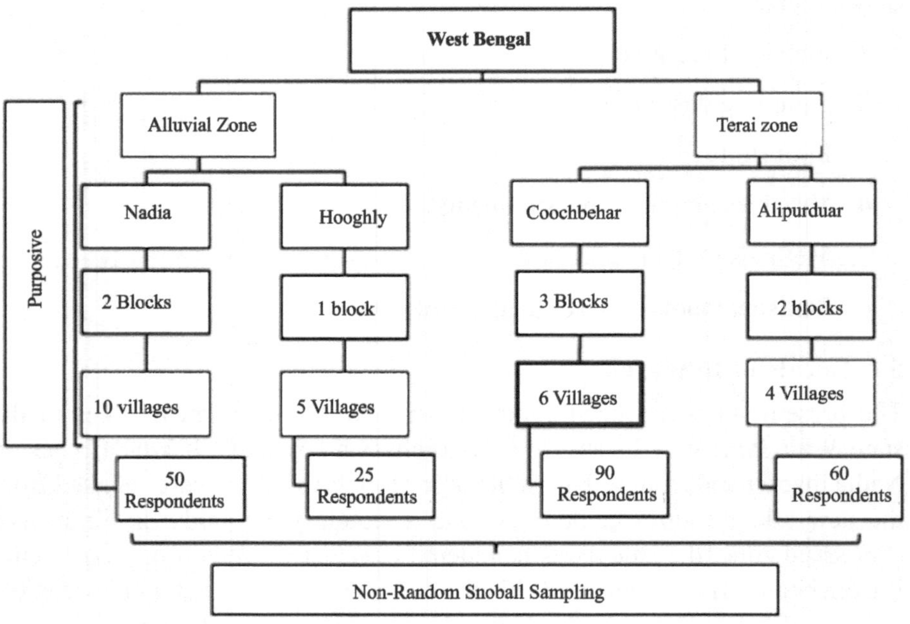

Fig. Schematic diagram of sampling method followed for the study

5.2. Sampling Design

Level	Location	Approach	Justification
State (1)	West Bengal	Purposive	Because of the limitation of time and resources.
Zone (2)	New Alluvial Zone (NAZ)	Purposive	Highly fertile lands and intensive agricultural lands poses threats to soil health declination. That is why the area needs more sustainable approaches like conservation agriculture.
	Terai Zone (TZ)		An already established CA platform for almost a decade which helps to isolate factors that helps in socialization of CA
District (4)	Nadia (NAZ)	Purposive	High cropping intensity and chances of high natural resource degradation in the region
	Hooghly (NAZ)		
	Coochbehar (TZ)		Have more area under CA as compared to the other districts of this zone
	Alipurduar (TZ)		
Block (8)	Haringhata	Purposive	These blocks are the most agriculturally advanced blocks experiencing high cropping intensity and degraded soil health
	Chakdaha		
	Balagarh		
	Coochbehar I		These blocks were the oldest blocks with CA practicing farmers
	Coochbehar II		
	Dinhata II		
	Alipurduar I		
	Falakata		
Villages (24)	NAZ – 14	Purposive	Only these were the villages were few farmers were practicing differential spectrum of CA
	TZ – 10		Advanced farmers with specialized trainings in CA.
Farmers (229)	NAZ -75	Snowballing	Since number of CA practitioners are still discrete and there was a pandemic, snowballing had seemed to be the best suited method
	TZ- 154		

5.3. Pilot Study

A pilot study has been conducted in the concerned area before constructing the data collecting devices. In course of this survey, informal discussion has been carried out with some farmers, local leaders, extension agencies and state agricultural officials of the localities. An outline of the socio-economic background of the farmers of the concerned villages, their interest in sustainable agricultural practices, experiences of sustainable farming and management strategies against the adversities helped in preparing and finalizing the interview schedule.

5.4. Variables and Their Measurement

There are two main categories of variables in the present study

a. Independent variables

b. Dependent variables

a. Independent variables

Independent Variables	Empirical Measurements
Age of the respondent (x_1)	Chronological age in years
Education (x_2)	Through scoring
Family size (x_3)	In number
Farm size (x_4)	In bigha
Cropping intensity (x_5)	In percentage
Number of land fragments (x_6)	In number
Annual income (x_7)	In Indian rupees
Income per capita (x_8)	In Indian rupees
Income per unit of land (x_9)	In Indian rupees
Annual expenditure (x_{10})	In Indian rupees
Stubble height (x_{11})	In inches
Volume of residue (x_{12})	In kilograms
Scientific orientation (x_{13})	Modified scale of Supe (2007)
Innovativeness (x_{14})	Modified scale of Singh (1997)
Extension agency contact (x_{15})	Scale developed by Nirban (2003)
Information seeking behavior (x_{16})	Modified scale of Bhairamkar (2009)
Residue management score (x_{17})	Through scoring
Perception on natural resource degradation (x_{18})	Through scoring
Number of livestock (x_{19})	In number
Mass media utilization (x_{20})	Modified scale of Meti (1990) and Kirankumari (1991)

Age (x_1)

Age is one of the most important determinants of social status and social role of the individual. Here age refers to the chronological years completed by the respondents at the time of interview. It has been recorded in integers.

Education (x_2)

Education can be operationalized as the level of formal schooling attained/ literacy acquired by the respondents at the time of interview. Education plays an important role in changing the behavior and outlook of an individual. It provides knowledge and skill to negotiate the dynamics of social life and grant a unique status in the society. Here, level of education has been categorized and scored as per followed:

Categories	Score
Illiterate	1
Primary education (upto class 4)	2
Middle school (upto class 8)	3
High school (upto class 10)	4
Intermediate (upto class 12)	5
Graduation & above	6

Family size (x_3)

Family size has been operationalized as the number of members in an individual's family. In this study, only those members have been considered who were staying together and were sharing meal in one chullah.

Farm size (x_4)

Farm size is measure of farm business. Operationally farm size may be defined as a tract of land possessed by an individual for the purpose of growing crops. In the present study, farming land owned by a family is taken as the measure of farm size.

Cropping intensity (x_5)

It has been operationalized as the proportion of total annual cropped area by the respondents to their size of operational holdings expressed in percentage. The cropping intensity of an individual farmer has been calculated by using the following formula:

$$\text{Cropping Intensity} = \frac{\text{Gross Cropped Area}}{\text{Net Cropped Area}} \times 100$$

Number of land fragments (x_6)

Land fragmentation is a prime character in South East Asian countries. Cultivation of fragmented lands often ends up to be energy and cost prodigal. In this study, it has been operationalized as disintegration of the total holding into number of pieces. That is, whether farmers' land is consolidated or disintegrated, if yes, then total number of disintegrated pieces is taken into account for this study.

Annual income (x_7)

For this study, annual income has been operationalized as the total annual income of the family earned through all the sources and various enterprises in terms of INR.

Income per capita (x_8)

Income per capita is a measure of economic well-being of a household. In this present study, income per capita has been conceptualized as annual income distributed equally among all the members of the family. It has been calculated using the following formula:

Income per capita = Annual Income (AI) / Family size

Income per unit of land (x_9)

Income per unit of land is also an economic measure to understand the economic status of a household. In this present study, income per unit of land has been conceptualized as income incurred from per bigha of land under cultivation. It has been calculated using the following formula:

Income per unit of land = Annual income (AI) / Total area under cultivation

Annual expenditure (x_{10})

Annual expenditure is also an economic measure to determine the economic variability between the households. It has been conceptualized as the total amount of money spend by all the family members in an year. This is expressed in terms of INR.

Stubble height (x_{11})

Stubble is the portion of stalk left at field above which the main crop is harvested. It has been conceptualized as the length of stalk above which the crop is harvested by the farmer. This length vary from farmer to farmer depending on the package of practices followed and method of harvesting. In this study, stubble height of only rice crop has been considered. The height has been measured in inches.

Volume of residue (x_{12})

Residues are leftovers in the field after the main crop has been harvested. It is generally of two types – field residue and processed residue. A large portion of nutrients remains deposited in the residues of the plants and thus is very important organic source of nutrients. In the present study, we have conceptualized volume of residue as the amount residue of rice crop generated on field. Only rice crop has been considered for this study. The amount of rice residues has been expressed in kilograms or kgs.

Scientific orientation (x_{13})

Scientific orientation refers to the extent to which an individual respondent is oriented towards the scientific background of the farming techniques while making decisions in farming and implementing them. A scale developed by Supe (2007) was used for measuring the variable with some little modification. The original scale consists of six statements including both negative as well as positive statements measured on a five point continuum. The scoring was done by assigning 0 for strongly disagree and 4 for strongly agree in positive statements and assigning 0 for strongly agree and 4 for strongly disagree in negative statements. The total score was calculated by summing up the scores of all the five statements.

Innovativeness (x_{14})

Innovativeness refers to the behaviour pattern of an individual who has interest and desire to seek changes in farming techniques and ready to introduce such changes into his operations when practical and feasible. The scale developed by Singh (1997) was used for the present study. The original scale consists of five statements measured on a five point continuum. The scoring was done by assigning 0 for strongly disagree and 4 for strongly agree in positive statements and assigning 0 for strongly agree and 4 for strongly disagree in negative statements. The total score was calculated by summing up the scores of all the five statements.

Extension agency contact (x_{15})

It has been operationalized as the frequency with which the local extension agents are in contact with respondents. In the present study we used a scale developed by Nirban (2004) with some modification. There is a list of extension agents and their contact frequency has been scored as once a month (3), once in two months (2), once in three months (1) and never (0). The score is given each of the extension agents operating in the locality which is then added to obtain the total score. This total score obtained is used in the study.

Information seeking behavior (x_{16})

It has been operationalized as the frequency with which the respondents approaches different sources to seek information. Information seeking behavior was measured for both the formal and informal sources separately. It was measured using a scale developed by Bhairamkar (2009) with a little modification. In this scale frequency was studied in a three point continuum that is regularly, occasionally and never and the numerical values of 2, 1 and 0 were assigned respectively.

Residue management score (x_{17})

Crop residues are managed by the farmers through a number of ways. In the present study we have selected few common practices of crop residue management. Total nine common practices were studied, viz., - stubble burning, for cattle feed, for fuel, as compost, retention for conservation agriculture, for thatching and fencing purpose, as mulch, for mushroom production and selling out. These practices were then scored on a 4 point continuum scale – mostly (2), sometimes (2), rare (1) and never (0). In case of non-sustainable practices the scoring was, always (0), rare (1), sometimes (2) and never (3). These scores obtained against each practices were then summed up and the total score was used in the final study.

Perception on natural resource degradation (x_{18})

Erosion and contamination of natural resources like soil and water is very common in present situation. To stop this degradation, perception development is a necessity. In the present study the respondents were given four broad domains viz., soil erosion, groundwater depletion, biodiversity erosion and agricultural pollution and were asked to score each of them out of 100 as per their perceived changes over the years. The scores against each domain were then added and mean value was calculated. This mean value was then used as a score against perception on natural resource degradation in the present study.

Number of livestock (x_{19})

Owning livestock is a common trait of rural India. In this study, only cattle has been considered as livestock. It has been operationalized as total number of cattle owned by a household.

Mass media utilization (x_{20})

It is operationalized as the frequency with which the mass media such as television, radio, newspaper, journals etc. are used by the respondents. The respondents were asked to indicate whether he/she has owned the media or not. The respondents were then also asked to indicate their degree of participation in terms of listening habit, viewing habit and reading habit. Scale developed by Meti (1990) and Kirankumari (1991) was used to measure this aspect. The subscribers and non-subscribers of the media were given the numerical value 1 and 0. The listening, viewing or reading habit was studied on a three point continuum i.e. regular, occasional and never and the numerical values of 2, 1 and 0 were assigned respectively.

b. Dependent variables

Perception on water management (y_1)

It has been conceptualized as the orientation of farmer towards conserving water resources and making efficient use of water. Water is an essential commodity not only in agriculture but also in the daily life. The respondents were provided four items on water management and were asked to rate them on a 4 point continuum scale developed through expert's opinion and scale construction method. The response was recorded as – more relevant (4), less relevant (3), somewhat relevant (2) and irrelevant (1). The score obtained against each statement are then added and the final score has been used to quantify perception on water management.

Perception on nutrient management (y_2)

It has been conceptualized as the orientation of farmer towards making efficient and judicious application of nutrients in their field. Balanced and proper nutrient application is an important input for plant growth and production. The respondents were provided four statements on nutrient management and were asked to rate them on a 4 point continuum scale developed through expert's opinion and scale construction method. The response was recorded as – more relevant (4), less relevant (3), somewhat relevant (2) and irrelevant (1). The score obtained against each statement are then added and the final score has been used to quantify perception on nutrient management.

Perception on carbon management (y_3)

It has been conceptualized as orientation of farmer towards maintenance and managing soil organic carbon. Organic carbon plays a vital role to upkeep soil health. The respondents were provided four items on carbon management and were asked to rate them on a 4 point continuum scale developed through expert's opinion and scale construction method. The response was recorded as – more relevant (4), less relevant (3), somewhat relevant (2) and irrelevant (1). The score obtained against each statement are then added and the final score has been used to quantify perception on carbon management.

Perception on ecosystem management (y_4)

It has been conceptualized as the orientation of farmer towards conserving biotic and abiotic components of the ecosystem. In the present study, perception on ecosystem management has been quantified by the score obtained by cube rooting the product score of perception on water management (y_1), perception on nutrient management (y_2) and perception on carbon management (y_3). The obtained score was used for the study. The formula used,

$y_4 = (y_1 * y_2 * y_3)^{1/3}$

5.5. Methods Used for Data Collection

Following methods were used for collecting primary data from the respondents:

a. **Preparation of interview schedule:** Interview schedule was used as the primary tool for data collection. It may be defined as 'a formal list, a category or inventory and it may be added that it is a counting device used in formal and standardized inquiries, the sole purpose of which is aiding the collection of quantitative and qualitative cross sectional data'. Comprehensive and detailed interview schedule consisting of structured as well as open ended questions was developed for collecting relevant information from the respondents.

b. **Pre-testing of interview schedule:** The structured interview schedule was pre tested on a non-sampled area and necessary corrections were made. The interview schedule was finalized after making necessary modification, deletion and addition based on pre testing recommendations.

c. **Scale construction for perception study:** A list of items for quantifying the perception of farmers has been finalized with the help of the responses received from the experts of concerned domains. At first a list of 21 items was prepared and experts were asked to rate them on a 4 point summated rating scale, based on their relevance to the concerned heads. The highest score (4) was assigned to 'most relevant' and lowest score (1) to 'least relevant' for each items. Total score of each item was calculated and four highest scored items were then selected for each of the 3 dependent variables, viz. perception on water management, perception on nutrient management and perception on carbon management. Responses were then marked and then four items for each head, i.e., Water Management, Nutrient Management and Carbon Management. Total 12 items are selected.

5.6. Statistical Analysis and Interpretation of Data (Analytic tools)

The data collected in the interview schedule have been processed and analyzed in accordance with the outline laid down for the purpose at the time of developing the research plan. Processing implies editing, coding, classification and tabulation of collected data. The statistical techniques and tools used in the present study are:

Mean

It is a measure of central tendency (or statistical averages) tells us the point about which items have a tendency to cluster. Such a measure is considered as the most representative figure for the entire mass of data. Measure of central tendency is also known as statistical average. Mean, median and mode are the most popular averages. Mean, also known as arithmetic average, is the most common measure of central tendency and may be defined as the value, which we get by dividing the total of the values of various given items in a series by the total number of items. We can work it out as follows

Mean, $\overline{x} = \frac{\sum x_{ij}}{N}$

Mean is the simplest measurement of central tendency and is a widely used measure. Its chief use consists in summarizing the essential features of a series and in enabling data to be compared. It is a relatively stable measure of central tendency, but it suffers from some limitations of unduly affected by extremes; it may not coincide with actual value of an item in a series, and it may lead to strong impressions, particularly when the item values are not given with the average. However, mean is better than other average, especially in economic and social studies where direct quantitative measurements are possible.

Standard Deviation

Standard deviation is the most widely used measure of dispersion of a series and is commonly denoted by the symbol σ (pronounced as sigma). Standard deviation is the square root of the arithmetic mean of the square of deviations, the deviations being measured from the arithmetic mean of distribution. It is less affected by sampling errors and is more stable measure of dispersion. It is worked out as follows,

Standard deviation, $\sigma = \frac{\sqrt{\Sigma(x-\overline{x})}}{N-1}$

Coefficient of Variation

A measure of variation which is independent of the unit of measurement is provided by coefficient of variation. Being unit free, this is useful for computation of variability between different populations. The coefficient of variation is standard deviation expressed as percentage of the mean and is measured by the formula.

Coefficient of correlation

When an increase or decrease in one variable is accompanied by an increase or decrease in other variable, the two are said to be correlated and the phenomenon is known as correlation. Correlation coefficient (r) is a measure of the relationship between two variables, which are at the interval or ratio level of measurement and are linearly related. A Spearman's coefficient of correlate

The value of 'r' lies between +1 to -1. Positive values of r indicate that positive correlation between the two variables (i.e. change in both variables takes place in the same direction), whereas negative values of 'r' indicate negative correlation i.e. changes in the two variables taking place in opposite direction. A zero value of 'r' indicates that there is no association between the two variables. When 'r' (+) 1, it indicates perfect positive correlation and when it is (-) 1, it indicates perfect negative correlation, meaning thereby that variations in independent variable (x) explain 100 per cent of the variations in the dependent variable (y). We can also say that for a unit change in independent variable, if there happens to be constant change in the dependent variable in the same direction, the correlation will be termed as perfect positive. But if such change occurs in the opposite direction, the correlation will be termed as perfect negative. The values of 'r' nearer to =1 or -1 indicates high degree of correlation between the two variables.

Stepwise Multiple Regression

Stepwise regression is a variation of multiple regressions which provides a means of choosing independent variables that in the best prediction possible with the fewest independent variables. It permits the user to solve a sequence of one or more multiple linear regression problems by stepwise application of the least square method. At each step in the analysis, a variable is added or removed which results in the greatest production of error sum of square (Burroghs corporation, 1975).

According to Drapper and Smith (1981), the method of stepwise multiple regression analysis is to insert variables in turn until the regression equation is satisfactory. The order of insertion is determined by using the partial correlation coefficient as the measure of the importance of variables not yet in the equation.

The program, according to Burrogh's Corporation (1975), first forms a correlation matrix, finds the best predictor (the independent variable having the highest correlation with criterion variable) and performs a regression analysis with this predictor. Then, the second best predictor (independent), and so on. At any given stage, the group of predictors being used is not necessarily the best group of that size (i.e. the particular group of independent variables does not necessarily have the highest multiple correlation with the criterion that any group of this size does). Rather, this group contains the variables that have the highest individual correlation with the criterion.

Stepwise regression is the step-by-step iterative construction of a regression model that involves the selection of independent variables to be used in a final model. It involves adding or removing potential explanatory variables in succession and testing for statistical significance after each iteration. Stepwise regression is a method that iteratively examines the statistical significance of each independent variable in a linear regression model.

The underlying goal of stepwise regression is, through a series of tests (e.g. F-tests, t-tests) to find a set of independent variables that significantly influence the dependent variable. Stepwise regression can be achieved either by trying out one independent variable at a time and including it in the regression model if it is statistically significant or by including all potential independent variables in the model and eliminating those that are not statistically significant. There are three approaches to stepwise regression:

1. **Forward selection-** It begins with no variables in the model, tests each variable as it is added to the model, then keeps those that are deemed most statistically significant—repeating the process until the results are optimal.
2. **Backward elimination-** This starts with a set of independent variables, deleting one at a time, then testing to see if the removed variable is statistically significant.
3. **Bidirectional elimination-** This is a combination of the first two methods that test which variables should be included or excluded.

Path Analysis

The term 'path analysis' was first introduced by the biologist, Sewall Wright in 1934 in connection with decomposing the total correlation between any two variables in a causal system. The technique of path analysis is based on a series of multiple regression analysis with the added assumption of causal relationship between independent and dependent variables. Path analysis makes use of standardized partial regression coefficient (known as beta weights) as effect coefficients. In linear additive effects are assumed, and then through path analysis a simple set of equations can be built up showing how each variable depends on preceding variables. The main principle of path analysis is that any correlation coefficient between two variables, or a gross or overall measure of empirical relationship can be decomposed into a series of paths: separate paths of influence leading through chronologically intermediate variable to which both the correlated variables have linked.

The merit of path analysis in comparison to correlation analysis is that it makes possible the assessment of the relative influence of each antecedent or explanatory variable on the consequent or criterion variables by first making explicit the assumptions underlying the causal connections and then by elucidating the indirect effect of the explanatory variables.

Canonical Covariate Analysis

A canonical correlation is the correlation of two canonical variables, one representing a set of independent variables, the other a set of dependent variables. Each set may be considered a latent variable based on measured indicator variables in its set. The canonical correlation is optimized such that the linear correlation between the two latent variables is maximized. Canonical correlation is used for many to many relationships. There may be more than one such linear correlation relating the two set of variables, with each such correlation representing a different dimension by which the independent set of variables is related to the dependent set. The purpose of canonical correlation is to explain the relation of the two sets of variables not to model the individual variables. In addition to asking how strong the relationship is between two latent variables, canonical correlation is useful in determining how many dimensions are needed to account for that relationship. Canonical correlation finds the linear combination of variables that produces the largest correlation with second set of variables.

Canonical correlation is a member of multiple general linear hypothesis (MLGH) family and shares many of assumptions of multiple regression such as linearity of relationship, homoscedasticity (same level of relationship for the full range of data), interval or near interval data, untruncated variables, proper specification of model, lack of high multicollinearity and multivariate normality for purpose of hypothesis testing.

Cluster Analysis

The purpose of cluster analysis is to reduce a large data set to meaningful subgroups of individuals or objects. The division is accomplished on the basis of similarity of the objects across a set of specified characteristics. Outliers are a problem with this technique, often caused by too many irrelevant variables. The sample should be representative of the population, and it is desirable to have uncorrelated factors. There are three main clustering methods hierarchical, this is a tree like process appropriate for smaller data set; non-hierarchical, which requires specification of the number of clusters apriori; and a combination of both. There are four main rules for developing clusters – the clusters should be different, they should be reachable, they should be measurable and the clusters should be profitable (big enough to matter). This is a great tool for market segmentation.

Artificial Neural Network Analysis

ANNs are computing systems inspired by the biological neural network that constitute animal brains. Such systems learn (progressively improve performance on) tasks by considering examples, generally without task specific programming. An ANN is based on a collection of connected units or nodes called artificial neurons (analogous to biological neurons in an animal brain). Each connection (analogous to a synapse) between artificial neurons can transmit a signal from one to another. The artificial neuron that receives the signal can process it and then signal artificial neurons connected to it. In common ANN implementation, the signal at a connection between artificial neurons is a real number, and the output of each artificial neuron is calculated by a non-linear function of the sum of its inputs. Artificial neurons and connections typically have a weight that adjusts as learning proceeds. The weight increases or decreases the strength of the signal at a connection. Artificial neurons may have a threshold such that only if the aggregate signal crosses that threshold is the signal sent. Typically, artificial neurons are organized in layers. Different layers may perform different kind of transformations on their inputs. Signals travel from the first (input), to the last (output) layer, possibly after traversing the layers multiple times. The original goal of the ANN approach was to solve problems in the same way that a human brain would. Over time, attention focused on matching specific mental abilities, leading to deviations from biology. ANNs have been used on a variety of tasks, including computer vision, speech recognition, machine translation, social network filtering, playing board and video games and medical diagnosis.

6

Case Studies

Table 1: Descriptive statistics of the socio-economic and communication characteristics of the respondents in terms of Range (Minimum and Maximum), Mean, Standard Deviation (SD) and Coefficient of Variation (CV) of Alluvial Zone in West Bengal

Variables	Range		Mean	SD	CV (%)
	Minimum	Maximum			
Age of the respondents (x_1)	27.00	82.00	51.49	12.432	24.14
Education (x_2)	1.00	6.00	4.00	1.433	35.83
Family size (x_3)	2.00	22.00	5.60	2.852	50.93
Farm size (x_4)	2.00	40.00	9.73	7.522	77.29
Cropping intensity (x_5)	155.56	400.00	260.65	51.283	19.66
No. of fragments (x_6)	1.00	22.00	7.47	4.966	66.50
Annual income (x_7)	85000.00	1540000.00	239702.00	229633.358	95.80
Income per capita (x_8)	16000.00	366250.00	48281.97	43632.285	90.37
Income per unit of land (x_9)	8303.13	95014.29	30644.23	16949.445	55.31
Annual expenditure (x_{10})	55000.00	624500.00	167038.67	105804.377	63.34
Stubble height (x_{11})	1.00	8.50	5.46	1.935	35.43
Volume of residue (x_{12})	1024.00	19456.00	7042.61	5438.195	77.22
Scientific orientation (x_{13})	12.00	26.00	19.19	3.278	17.09
Innovativeness (x_{14})	7.00	19.00	14.36	2.305	16.05
Extension agency contact (x_{15})	2.00	9.00	4.91	1.678	34.20
Information seeking behavior (x_{16})	5.00	15.00	10.64	2.529	23.77
Residue management score (x_{17})	4.00	19.00	11.85	3.420	28.85
Perception on natural resource degradation (x_{18})	33.75	87.50	61.46	12.616	20.53
No. of livestock (x_{19})	1.00	6.00	1.81	1.099	60.61
Mass media utilization (x_{20})	2.00	8.00	5.52	1.554	28.15
Perception on water management in CA (y_1)	4.00	16.00	13.40	3.029	22.60
Perception on nutrient management in CA (y_2)	9.00	16.00	14.01	1.653	11.80
Perception on carbon management in CA (y_3)	6.00	16.00	12.11	2.463	20.35
Perception on ecosystem management in CA (y_4)	7.23	15.66	13.03	1.955	15.01

Table 1 Represents the distribution of variables in terms of range i.e. maximum and minimum values, mean, standard deviation and coefficient of variation for alluvial zone of West Bengal.

It has been found that for the independent variable **age (x_1)**, the range lies between 82 and 27 years, where the oldest respondent is of 82 years and the youngest respondent being 27years old. The mean value of the age of the respondents in this study was found to be 51.49 years with a standard deviation of 12.42 in the dataset. Coefficient of variation for this variable was found to be 24.14 per cent, which indicates a high level of consistency in the distribution.

It has been found that for the independent variable **education (x_2)**, the range lies between the scale value 1 and 6, which means the minimum qualification amongst the respondents is illiteracy and the highest formal education received among the respondents is either a graduate or post graduate. The mean scale value lies at 4, which means majority of the respondents received formal education upto high school. The standard deviation in the data set is measured at 1.43. Coefficient of variation for this variable was found to be 35.83 per cent, which indicates the level of consistency in the distribution is at higher end.

It has been found that for the independent variable **family size (x_3)**, the range stretches within a minimum of 2 and a maximum of 22 family members. The average size of family consists of approximately 6 (5.60) members. The measured standard deviation in the dataset was found to be 2.85 and the calculated coefficient of variation is 50.93 per cent. This indicates there is a moderate to high level of consistency in distribution of data.

It has been found that for the independent variable **Farm size (x_4)**, the range stretches between minimum of 2 bighas and maximum of 40 bighas of land. The mean value of farm size of the respondents in this study was found to be 9.73 bighas with a standard deviation of 7.52 in the dataset. Coefficient of variation for this variable was found to be 77.29 per cent, which indicates a moderate level of consistency in the distribution.

It has been found that for the independent variable **cropping intensity (x_5)**, the minimum was observed at 155.56 per cent and maximum at 400.00 per cent. The mean value of cropping intensity in this study stands at 260.65 per cent with a standard deviation of 51.28 in the dataset. Coefficient of variation observed for this variable was 19.68 per cent, which indicates a high level of consistency in the distribution.

It has been found that for the independent variable **number of land fragments (x_6)**, the range lies between 1 (minimum) and 22 (maximum). The mean value of number of land fragments observed in this study was 7.47 years with a standard deviation of 4.97. Coefficient of variation for this variable was 66.50 per cent, which indicates a moderate level of consistency in the distribution.

It has been found that for the independent variable **annual income (x_7)**, the minimum is 85000 INR and the maximum is 1540000 INR. The mean value is 239702.00 INR with a standard deviation of 229633.35 in the dataset. Coefficient of variation for this variable was found to be 95.80 per cent, which indicates a low level of consistency in the distribution.

It has been found that for the independent variable **income per capita (x_8)**, the minimum is 16000 INR and the maximum is 366250 INR. The mean value is 48281.97 INR and the standard deviation is 43632.29. Coefficient of variation for this variable was found to be 90.37 per cent, which indicates a low level of consistency in the distribution.

For the independent variable **income per unit of land (x_9)**, it has been found the minimum is 8303.13 INR and the maximum is 95014.29 INR. The mean value is 30644.23 with a standard deviation of 16949.45 in the total distribution. Coefficient of variation for this variable was found to be 55.31 per cent, which indicates a medium level of consistency in the distribution.

It has been found that for the independent variable **annual expenditure (x_{10})**, the minimum expenditure is 55000 INR and the maximum is 624500 INR. The mean value is 167038.67 INR with a standard deviation of 105804.38 in the total distribution. Coefficient of variation estimated at 63.34 per cent, which indicates a medium level of consistency in the distribution.

It has been found that for the independent variable **stubble height (x_{11})**, the minimum height maintained is 1 inch and the maximum is 8.5 inches. The mean height is 5.46 inches with a standard deviation of 1.94 in total distribution. Coefficient of variation for this variable was found to be 35.43 per cent, which indicates a moderate level of consistency in the distribution.

It has been found that for the independent variable **volume of residue (x_{12})**, the minimum amount of residue generated in a farmer's field was 1024 kgs and the maximum amount was 19456 kgs. The mean amount was 7042.61 with a standard deviation of 5438.20 in the total distribution. Coefficient of variation for this variable was 77.22 per cent, which indicates a low level of inconsistency in the distribution.

It has been found that for the independent variable **scientific orientation (x_{13})**, the minimum score is 12 and the maximum score obtained is 26. The mean score lies at 19.19 with a standard deviation of 3.28 in the total distribution. Coefficient of variation for this variable was found to be 17.09 per cent, which indicates a very high level of consistency in the distribution.

It has been found that for the independent variable **innovativeness (x_{14})**, the minimum score is 7 and the maximum score obtained is 19. The mean score lies at 14.36 with a standard deviation of 2.31 in the total distribution. Coefficient of variation for this variable was found to be 16.05 per cent, which indicates a very high level of consistency in the distribution.

It has been found that for the independent variable **extension agency contact (x_{15})**, the minimum score is 2 and the maximum score obtained is 9. The mean score lies at 4.91 with a standard deviation of 1.68 in the total distribution. Coefficient of variation for this variable was found to be 34.20 per cent, which indicates a moderately high level of consistency in the distribution.

It has been found that for the independent variable **information seeking behaviour (x_{16})**, the minimum score is 5 and the maximum score obtained is 15. The mean score lies at 10.64 with a standard deviation of 2.53 in the total distribution. Coefficient of variation for this variable was found to be 23.77 per cent, which indicates a high level of consistency in the distribution.

It has been found that for the independent variable **residue management score (x_{17})**, the minimum score is 4 and the maximum score obtained is 19. The mean score lies at 11.85 with a standard deviation of 3.42 in the total distribution. Coefficient of variation for this variable was found to be 28.85 per cent, which indicates a high level of consistency in the distribution.

It has been found that for the independent variable **perception on natural resource degradation (x18)**, the minimum score is 33.75 and the maximum score obtained is 87.50. The mean score lies at 61.46 with a standard deviation of 12.62 in the total distribution. Coefficient of variation for this variable was found to be 20.53 per cent, which indicates a high level of consistency in the distribution.

It has been found that for the independent variable **number of livestock owned (x_{19})**, the minimum count is 1 and the maximum is 6. The mean count is approximately 2 (1.81) with a standard deviation of 1.10 in the total distribution. Coefficient of variation for this variable was found to be 60.61 per cent, which indicates a medium level of consistency in the distribution.

It has been found that for the independent variable **mass media utilization (x_{20})**, the minimum score is 2 and the maximum score obtained is 8. The mean score lies at 5.52 with a standard deviation of 1.55 in the total distribution. Coefficient of variation for this variable was found to be 28.15 per cent, which indicates a high level of consistency in the distribution.

It has been found that for the dependent variable **perception on water management in CA (y_1)**, the minimum score is 4 and the maximum score obtained is 16. The mean score lies at 13.40 with a standard deviation of 3.03 in the total distribution. Coefficient of variation for this variable was found to be 22.60 per cent, which indicates a high level of consistency in the distribution.

It has been found that for the dependent variable **perception on nutrient management in CA (y_2)**, the minimum score is 9 and the maximum score obtained is 16. The mean score lies at 14.01 with a standard deviation of 1.65 in the total distribution. Coefficient of variation for this variable was found to be 11.80 per cent, which indicates a very high level of consistency in the distribution.

It has been found that for the dependent variable **perception on carbon management in CA (y_3)**, the minimum score is 6 and the maximum score obtained is 16. The mean score lies at 12.11 with a standard deviation of 2.46 in the total distribution. Coefficient of variation for this variable was found to be 20.35 per cent, which indicates a high level of consistency in the distribution.

It has been found that for the dependent variable **perception on ecosystem management in CA (y_4)**, the minimum score is 7.23 and the maximum score obtained is 15.66. The mean score lies at 13.03 with a standard deviation of 1.96 in the total distribution. Coefficient of variation for this variable was found to be 15.01 per cent, which indicates a very high level of consistency in the distribution.

Table 2: Correlation Coefficient of perception on water management (y_1) vs. 20 independent variables (x_1-x_{20}) in New Alluvial Zone of West Bengal (n=75)

Independent Variables	'r' Value	Remarks
Age of the respondents (x_1)	0.448	**
Education (x_2)	0.192	
Family size (x_3)	-0.171	
Farm size (x_4)	0.153	
Cropping intensity (x_5)	-0.088	
No. of fragments (x_6)	-0.426	**
Annual income (x_7)	0.044	
Income per capita (x_8)	0.143	
Income per unit of land (x_9)	-0.434	**

Independent Variables	'r' Value	Remarks
Annual expenditure (x_{10})	-0.062	
Stubble height (x_{11})	0.480	**
Volume of residue (x_{12})	0.140	
Scientific orientation (x_{13})	0.014	
Innovativeness (x_{14})	0.104	
Extension agency contact (x_{15})	0.150	
Information seeking behavior (x_{16})	0.157	
Residue management score (x_{17})	-0.055	
Perception on natural resource degradation (x_{18})	0.083	
No. of livestock (x_{19})	0.140	
Mass media utilization (x_{20})	-0.030	

**Correlation is significant at the 0.01 level
*Correlation is significant at the 0.05 level

Table 2 presents the correlation coefficient of correlation between perception on water management (y_1) and 20 independent variables (x_1-x_{20}). It is discernable from the table that the variables, **age of the respondent** and **stubble height retention** have recorded **positive and significant** correlation, whereas **number of fragments** and **income per unit of land** have recorded **negative but significant** correlation with the dependent variable under discussion.

The results suggest that the higher is the age of the farmers, the better has been the water management. This is because aged farmers are more experienced about water requirement and water stress condition and how to overcome that strategically through managing the resource. The negative correlation with income per unit of land reveals that, the perception on water management has failed to make any positive contribution on income of the farmers in the alluvial zone. Fragmentation hinders adoption of efficient water management since it makes farms more cost prodigal. Since this area enjoys a rainfall regime ranging from 1700-1900mm per year, farmers do not pay additional attention for water management in their farms. This offers a real challenge for the socialization of Conservation Agriculture in terms of water management wherein the farmers are already blessed with bounty of water, both in terms of ground water regime and total annual precipitation.

Table 3: Path analysis of perception on water management (y_1) vs. 20 exogenous variables in New Alluvial Zone of West Bengal

Variables	Total Effect (TE)	Direct Effect (DE)	Indirect Effect (IE)	Highest Indirect Effect (HIE)
Age of the respondents (x_1)	0.448	0.268	0.180	0.087(x_6)
Education (x_2)	0.192	0.002	0.190	0.050(x_9)
Family size (x_3)	-0.171	-0.117	-0.054	-0.088(x_6)
Farm size (x_4)	0.153	0.071	0.082	0.101(x_9)
Cropping intensity (x_5)	-0.088	0.005	-0.093	-0.066(x_9)
No. of fragments (x_6)	-0.426	-0.312	-0.114	-0.075(x_1)
Annual income (x_7)	0.044	0.023	0.021	-0.065(x_8)
Income per capita (x_8)	0.143	-0.115	0.258	0.076(x_{10})
Income per unit of land (x_9)	-0.434	-0.269	-0.165	-0.105(x_{11})
Annual expenditure (x_{10})	-0.062	0.131	-0.193	-0.070(x_3)
Stubble height (x_{11})	0.480	0.319	0.161	0.089(x_9)
Volume of residue (x_{12})	0.140	0.023	0.117	0.073(x_9)
Scientific orientation (x_{13})	0.014	-0.151	0.165	0.066(x_9)
Innovativeness (x_{14})	0.104	-0.011	0.115	0.033(x_{11})
Extension agency contact (x_{15})	0.150	0.111	0.039	0.025(x_9)
Information seeking behavior (x_{16})	0.157	0.093	0.064	0.092(x_9)
Residue management score (x_{17})	-0.055	0.010	-0.065	-0.097(x_{11})
Perception on natural resource degradation (x_{18})	0.083	-0.041	0.124	0.041(x_{19})
No. of livestock (x_{19})	0.140	0.114	0.026	0.034(x_{11})
Mass media utilization (x_{20})	-0.030	0.047	-0.077	-0.056(x_6)

Residual effect: 0.428

Table 3 presents the path analysis of dependent variable perception on water management (y_1) wherein the total effect (coefficient of correlation) has been decomposed into direct, indirect and residual effect.

It has been recorded that the variable stubble height retention has exerted the highest direct effect on perception on water management (y_1). It implies that for alluvial agro ecosystem, water management perception has got a collateral impact exerted by stubble height retention. This is because stubbles help to conserve soil moisture by covering the ground surface and also reduces evaporation loss. This curb down the water requirement of the next crop, especially in the sowing time. Thus stubble height retained in the field by the farmers has direct implications on their water management strategies.

The highest indirect effect has been exerted by exogenous variable income per capita. This is extremely important that income has been a strong determinant in deciding the extent of water management by the farmers. The research continuously points out that without consideration of the economy of CA, the ecology of CA cannot be addressed.

The variable income per unit of land has routed the highest indirect effect in as many as eight variables to ultimately characterize and scale up the level of water management.

The residual effect being 0.428 means 42.8% of the variance could not explained by the set of exogenous variables.

Table 4: Stepwise regression analysis of perception on water management (y_1) versus 20 independent ($x_1 - x_{20}$) variables in New Alluvial zone of West Bengal (n=75)

Variables	Reg. coeff. B	S.E. B	Beta	t value	R^2	Std. error of the estimate
Stubble height (x_{11})	0.451	0.143	0.286	3.155	0.518	2.175
Age of the respondents (x_1)	0.067	0.022	0.273	3.072		
No. of fragments (x_6)	-0.186	0.055	-0.302	-3.368		
Income per unit of land (x_9)	0.000	0.000	-0.300	-3.298		

Table 4 shows that variables, stubble height retention (x_{11}), age of the respondent (x_1), number of fragments (x_6) and income per unit of land (x_9) have been retained at the last step. The R^2 value being 0.518, reveals that these four variables together explains 51.8 per cent of the variance embedded in perception on water management (y_1).

Stubble height is an important part for conserving soil moisture. When optimum amount of stubble is retained, it helps a farm to save on their water requirement; more during sowing season. Aged farmers with better experiences on managing water stress condition. Fragmentation of holding has got a deleterious impact on energy and cost management of a farm. With more fragmented land, the drudge of management will automatically increase, turning the farm energy and cost prodigal. This is the reason fragmentation as a functional variable is so important to predict water management. The results rightly direct the attention of the conservationists in agriculture into the need for upscaling return, that too into a happy return, through pursuing efficient and cost effective water management.

We can go increasingly optimistic about the commitment and participation of young generation farmers in this zone in favor of socializing CA beyond "illusion of energy and chemical input prodigal agriculture". The farm economy of CA has unabatedly being figure up to justify the need and essentiality for upscaling income, consistent and increasing from CA beyond scales and distribution.

Table 5: Correlation Coefficient of perception on nutrient management (y_2) vs. 20 independent variables (x_1-x_{20}) in New Alluvial Zone of West Bengal (n=75)

Independent Variables	'r' Value	Remarks
Age of the respondents (x_1)	0.101	
Education (x_2)	0.357	**
Family size (x_3)	-0.227	
Farm size (x_4)	0.110	
Cropping intensity (x_5)	-0.100	
No. of fragments (x_6)	-0.047	
Annual income (x_7)	0.218	
Income per capita (x_8)	0.203	
Income per unit of land (x_9)	-0.405	**
Annual expenditure (x_{10})	0.028	
Stubble height (x_{11})	0.318	**
Volume of residue (x_{12})	0.184	
Scientific orientation (x_{13})	0.161	
Innovativeness (x_{14})	0.506	**
Extension agency contact (x_{15})	-0.033	
Information seeking behavior (x_{16})	0.293	*
Residue management score (x_{17})	0.228	*
Perception on natural resource degradation (x_{18})	0.211	
No. of livestock (x_{19})	0.142	
Mass media utilization (x_{20})	0.070	

**Correlation is significant at the 0.01 level

*Correlation is significant at the 0.05 level

Table 5 presents the coefficient correlation between the perception of nutrient management (y_2) vs 20 independent variables. It has been found that the variables **stubble height retention, innovativeness, information seeking behavior and residue management score** have recorded **positive and significant** correlation with the dependent variable, perception on nutrient management (y_2) under study. The variable **income per unit of land** has shown a **negative but significant** correlation with the same dependent variable.

The variable residue management score added to build up better perception on better nutrient mgt. the upcoming of org farming and a substantive improvement in pursuing pulses and oilseed in their basket of choices have contributed to a better understanding on soil nutrient mgt. the inclusion of vegetable crops acts also added to their perception building on nutrient management alongwith management of crop residue.

The negative correlation between income per unit of land and perception on nutrient management can be relegated to the poor functioning or non-functioning of marketing services as expected to contribute to their per unit area either.

The text of the table 5 implies that for new alluvial zone with special reference to the research locale under study, indicates that, the improvement in water and soil nutrient management has not been defended by a happy return and it may offer a socio-ecological threats for socializing CA as per standards and values.

The learning experience- everything is fine and joyful for farmers investing their hard earned pennies to mentor their farming when there has been a happy return, may it be CA, ecological farming and even if it is a conventional way of farming. CA has to be economically sustainable, socially compatible and ofcourse ecologically rewarding in the form of better ecological health and resilience. It is really highly complicated proposition, may not be that impossible, to connect and correlate ecology with economy and vice-versa.

Table 6: Path analysis of perception on nutrient management (y_2) vs. 20 exogenous variables in New Alluvial Zone of West Bengal

Variables	**Total Effect**	**Direct Effect**	**Indirect Effect**	**Highest Indirect Effect**
Age of the respondents (x_1)	0.101	-0.056	0.157	0.046(x_9)
Education (x_2)	0.357	0.186	0.171	0.082(x_{10})
Family size (x_3)	-0.227	-0.261	0.034	0.202(x_{10})
Farm size (x_4)	0.110	-0.087	0.197	0.204(x_{10})
Cropping intensity (x_5)	-0.100	0.003	-0.103	-0.056(x_9)
No. of fragments (x_6)	-0.047	0.041	-0.088	-0.074(x_3)
Annual income (x_7)	0.218	0.095	0.123	0.124(x_{10})
Income per capita (x_8)	0.203	-0.168	0.371	0.196(x_{10})
Income per unit of land (x_9)	-0.405	-0.230	-0.175	-0.091(x_{11})
Annual expenditure (x_{10})	0.028	0.339	-0.311	-0.155(x_3)
Stubble height (x_{11})	0.318	0.275	0.043	-0.079(x_{17})
Volume of residue (x_{12})	0.184	0.067	0.117	0.062(x_9)
Scientific orientation (x_{13})	0.161	0.048	0.113	0.056(x_3)
Innovativeness (x_{14})	0.506	0.460	0.046	-0.048(x_3)
Extension agency contact (x_{15})	-0.033	-0.019	-0.014	0.074(x_{14})
Information seeking behavior (x_{16})	0.293	0.087	0.206	-0.079(x_9)
Residue management score (x_{17})	0.228	0.258	-0.030	0.084(x_{11})
Perception on natural resource degradation (x_{18})	0.211	0.057	0.154	0.059(x_{17})
No. of livestock (x_{19})	0.142	0.039	0.103	0.066(x_{14})
Mass media utilization (x_{20})	0.070	-0.121	0.191	0.156(x_{14})

Residual effect: 0.368

Table 6 presents the path analysis of dependent variable perception on nutrient management (y_2) wherein the total effect (coefficient of correlation) has been decomposed into direct, indirect and residual effect.

The variable **innovativeness** has exerted **highest direct effect**. Modern and appropriate nutrient management in today's agriculture demands more skill and knowledge as to how, when and why the fertilizers shall be applied. That is why the farmers need to acquire skills and knowledge in nutrient management. A disproportionate application of NPK not only deteriorates soil quality and crop productivity, it also yields into ecological fall out and coercion. That is why innovativeness has got a major effect in building perception on nutrient management.

The **highest indirect effect** has been exerted by exogenous variable **annual expenditure**. This is extremely important that expenditure has been a strong determinant in deciding the extent of nutrient management by the farmers. The research continuously points out that without consideration of the economy of CA, the ecology of CA cannot be addressed. Annual expenditure has also routed the highest individual indirect effect on as many as 5 exogenous variables.

The residual effect being 0.368 means 36.8% of the variance could not be explained by the set of exogenous variables.

Table 7: Stepwise regression analysis of perception on nutrient management (y_2) versus 20 independent ($x_1 - x_{20}$) variables in New Alluvial Zone of West Bengal (n=75)

Variables	Reg. coeff. B	S.E. B	Beta	t value	R^2	Std. error of the estimate
Innovativeness (x_{14})	0.326	0.059	0.452	5.545	0.551	1.155
Income per unit of land (x_9)	0.000	0.000	-0.209	-2.286		
Education (x_2)	0.292	0.096	0.251	3.038		
Residue management score (x_{17})	0.137	0.044	0.281	3.144		
Stubble height (x_{11})	0.226	0.080	0.263	2.809		

Table 7 shows that variable, innovativeness (x_{14}), income per unit of land (x_9), education (x_2), residue management score (x_{17}) and stubble height retention (x_{11}) has been retained in the last step. The R^2 value being 0.551, reveals that this variable alone explains 55.10 per cent of the variance embedded in perception on nutrient management (y_2).

The nutrient management for CA needs both pragmatism and innovation. While the mode and amount of chemical fertilizers need to undergo gradual rationalization, it should be complemented by gradual increase in the source of natural organic nutrients. The crop residue management and maintaining stubble height on farm soil should be reciprocally responded with the gradual withdrawal of coercive chemical fertilizers. This complex but ecological tuning process certainly needs ecological learning, environmental pedagogy and hands on skills. The need for innovation has been so conspicuous in the pandemic time wherein managing ecology and retrieval of its resilience are of immense importance. Ntshangase et al. (2018) also found in their study that education could be a potential determinant in the adoption of no-till CA. Other studies (Knowler and Bradshaw, 2007; Nyambose and Jumbe, 2013) have shown that higher education levels increase the chances of adopting no-till CA because educated farmers are more likely to easily understand and be receptive to new technology or innovations.

Table 8: Correlation Coefficient of perception on carbon management (y_3) vs. 20 independent variables (x_1-x_{20}) in New Alluvial Zone of West Bengal (n=75)

Independent Variables	'r' Value	Remarks
Age of the respondents (x_1)	0.231	*
Education (x_2)	0.008	
Family size (x_3)	-0.215	
Farm size (x_4)	0.038	
Cropping intensity (x_5)	-0.149	
No. of fragments (x_6)	0.058	
Annual income (x_7)	0.116	
Income per capita (x_8)	0.086	
Income per unit of land (x_9)	-0.375	**
Annual expenditure (x_{10})	-0.204	
Stubble height (x_{11})	0.084	
Volume of residue (x_{12})	0.369	**
Scientific orientation (x_{13})	0.342	**
Innovativeness (x_{14})	0.010	
Extension agency contact (x_{15})	0.217	
Information seeking behavior (x_{16})	0.487	**
Residue management score (x_{17})	0.190	
Perception on natural resource degradation (x_{18})	0.438	**
No. of livestock (x_{19})	0.176	
Mass media utilization (x_{20})	0.308	**

**Correlation is significant at the 0.01 level

*Correlation is significant at the 0.05 level

Table 8 delineates the correlation coefficient between the perception on carbon management (y_3) vs. 20 independent variables. It has been found that variables **age, volume of residue generated, scientific orientation, information seeking behavior, perception on natural resource degradation and mass media utilization** delineates a **positive** and significant relationship with **perception on carbon management** in this study. The variable **income per unit of land** again laid a **negative** but significant relationship with the dependent variable under this study.

Carbon management is a complex yet important thing that needs to be taken care of in maintaining the sustainability of the land in long run. To build up an effective perception on carbon management one needs to develop a scientific temperament. This scientific temperament is often developed through exposure to different cosmopolite sources of information including mass media. Perception on natural resource degradation helps to conserve natural resources like crop residue, which if managed properly, is a great source for not only nutrients but also organic carbon for the soil. Managing natural resources like crop residues needs proper and definite knowledge. Here also, information sources and mass media plays an important role. Thus the results rightly shows the association of the aforesaid variables with perception development on carbon management. This finding is line with other research finding which says information is an important predictor is adoption of new agricultural technologies (Jabbar et al., 2003).

In alluvial zone, time and again, income has come as a determinant factor in perception development. Farmers being mostly small and marginal with limited income from agriculture, mostly restrain from trying new technologies if income is not certain. This hinders the perception development of farmers. Benefits of carbon management is mostly invisible or takes a long time to reflect in the income difference of the farmers. This is why the negative relation reflected in the present study is justified.

Table 9: Path analysis of perception on carbon management (y_3) vs. 20 exogenous variables in New Alluvial Zone of West Bengal

Variables	Total Effect (TE)	Direct Effect (DE)	Indirect Effect (IE)	Highest Indirect Individual Effect (HIIE)
Age of the respondents (x_1)	0.231	0.140	0.091	**0.038(x_{18})**
Education (x_2)	0.008	-0.023	0.031	**0.040(x_{18})**
Family size (x_3)	-0.215	-0.059	-0.156	-0.049(x_4)
Farm size (x_4)	0.038	-0.096	0.134	0.056(x_9)

Variables	Total Effect (TE)	Direct Effect (DE)	Indirect Effect (IE)	Highest Indirect Individual Effect (HIIE)
Cropping intensity (x_5)	-0.149	-0.069	-0.080	-0.037(x_{13})
No. of fragments (x_6)	0.058	0.055	0.003	0.045(x_{20})
Annual income (x_7)	0.116	-0.087	0.203	**0.056(x_{18})**
Income per capita (x_8)	0.086	-0.026	0.112	-0.050(x_7)
Income per unit of land (x_9)	-0.375	-0.150	-0.225	-0.079(x_{16})
Annual expenditure (x_{10})	-0.204	0.063	-0.267	-0.070(x_{20})
Stubble height (x_{11})	0.084	-0.005	0.089	0.049(x_9)
Volume of residue (x_{12})	0.369	0.250	0.119	0.055(x_{16})
Scientific orientation (x_{13})	0.342	0.122	0.220	0.054(x_{12})
Innovativeness (x_{14})	0.010	-0.109	0.119	0.086(x_{20})
Extension agency contact (x_{15})	0.217	0.236	-0.019	-0.038(x_{20})
Information seeking behavior (x_{16})	0.487	0.231	0.256	0.061(x_{18})
Residue management score (x_{17})	0.190	0.008	0.182	0.060(x_{18})
Perception on natural resource degradation (x_{18})	0.438	0.266	0.172	0.053(x_{16})
No. of livestock (x_{19})	0.176	0.078	0.098	0.096(x_{18})
Mass media utilization (x_{20})	0.308	0.253	0.055	0.054(x_{18})

Residual effect: 0.408

Table 9 presents the path analysis of dependent variable, perception on carbon management (y_3), wherein the total effect (coefficient of correlation) has been decomposed into direct, indirect and residual effect.

The variable perception on natural resource degradation has exerted the highest direct effect on the dependent variable perception on carbon management (y_3). A large part of the organic carbon is added to the soil from outside from various natural resources like crop residues, farm yard manure etc. Thus, perception on natural resource degradation and their recycling is very likely to have a strong effect on the perception development on managing organic carbon at field level by the respondent farmers.

Here also the results revealed that annual expenditure has exerted highest indirect effect on the dependent variable perception on carbon management. Be it income or expenditure, farm economy directly or indirectly influence the perception development on the farmer. To manage soil organic carbon, one needs to invest in maintaining on farm residues, application of organic manure etc. All these involves an expenditure. If the return against this expenditure is not confirmed, the effort in building up perception on carbon management for future sustainability will turn futile.

The variable perception on natural resource degradation has also routed the highest indirect individual effect on as many as 7 exogenous variables.

The residual effect is 0.408, which means 40.8% of the variance remains unexplained by the set of 20 exogenous variables (x_1–x_{20}).

Table 10: Stepwise regression analysis of perception on carbon management (y_3) versus 20 independent ($x_1 - x_{20}$) variables in New Alluvial Zone of West Bengal (n=75)

Variables	Reg. coef. B	S.E. B	Beta	t value	R^2	Std. error of the estimate
Information seeking behavior (x_{16})	0.260	0.090	0.266	2.905	0.526	1.782
Perception on natural resource degradation (x_{18})	0.056	0.017	0.287	3.296		
Volume of residue (x_{12})	0.000	0.000	0.241	2.678		
Mass media utilization (x_{20})	0.378	0.142	0.237	2.662		
Extension agency contact (x_{15})	0.307	0.129	0.208	2.392		
Scientific orientation (x_{13})	0.141	0.067	0.187	2.095		

Table 10 shows that variables, information seeking behavior (x_{16}), perception on natural resource degradation (x_{18}), volume of residue generated (x_{12}), mass media utilization (x_{20}), extension agency contact (x_{15}) and scientific orientation (x_{13}) have been retained at the last step. The R^2 value being 0.526, reveals that these variables together explains 52.6 per cent of the variance embedded in perception on carbon management (y_3).

Carbon management is a complex process involving scientific understanding. This understanding has been built up through continuous access to different reliable and authentic information sources, mass media and different extension agents. Organic carbon is a natural resource and can be restored and replenished through conservation of natural resources like crop residues. These variables thus fit perfectly in predicting the perception on carbon management of the farmers.

Table 11: Correlation Coefficient of perception on ecosystem management in CA (y_4) vs. 20 independent variables (x_1-x_{20}) in New Alluvial Zone of West Bengal (n=75)

Independent Variables	'r' Value	Remarks
Age of the respondents (x_1)	0.390	**
Education (x_2)	0.190	
Family size (x_3)	-0.274	*
Farm size (x_4)	0.126	
Cropping intensity (x_5)	-0.132	
No. of fragments (x_6)	-0.240	*

Independent Variables	'r' Value	Remarks
Annual income (x_7)	0.131	
Income per capita (x_8)	0.173	
Income per unit of land (x_9)	-0.523	**
Annual expenditure (x_{10})	-0.141	
Stubble height (x_{11})	0.398	**
Volume of residue (x_{12})	0.291	*
Scientific orientation (x_{13})	0.211	
Innovativeness (x_{14})	0.205	
Extension agency contact (x_{15})	0.176	
Information seeking behavior (x_{16})	0.379	**
Residue management score (x_{17})	0.110	
Perception on natural resource degradation (x_{18})	0.297	**
No. of livestock (x_{19})	0.213	
Mass media utilization (x_{20})	0.143	

**Correlation is significant at the 0.01 level
*Correlation is significant at the 0.05 level

Table 11 presents the coefficient correlation between the perception on ecosystem management of CA (y_4) vs 20 independent variables. It has been found that the variables age, stubble height retention, volume of residue generated, information seeking behavior and perception on natural resource degradation have recorded a positive and significant correlation; whereas the variables family size, number of fragments and income per unit of land have shown a negative but significant correlation with the dependent variable under study. Family size is an important determining factor in decision making. Ecosystem is nothing but an interrelated mechanism of interaction between human and the surrounding environment. Proper and sustainable management of this ecosystem requires enough skill and knowledge in managing different components working together. With more family members, imparting quality education is a challenging task if income is not secured and consistent. This incomplete or scattered knowledge is a hindrance in perception building on ecosystem management on CA.

Land fragmentation is another important factor in ecosystem management. Most of farmers in the study area being small and marginal are constrained by small, fragmented holdings and access to farm resources. With limited resources, the farmers tried to make the optimum use of those resources. These farmers from their experiences over the years have developed an idea of optimum resource combination that would help them in better ecosystem management in their small holdings.

Tiny and fragmented holding offers serious ecological disruption whenever the conservation of resources are in concern. The more is the fragmentation, the more would be the wastage of resources which may lead to the prodigal nature of the farm management. It is not prodigal in terms of energy management, it would be prodigal in terms of resource management, labour management etc. More the fragmentation, higher will be the need for energy and human labour both. When the farmers are passing through these conflict and confrontation between ecology and economy of the small farming and fragmented farming, the higher would be the perception on the need for restoration of ecological erosion and degradation. That is why it is the need of the hour to bring all these owners of small and fragmented holding together with a view to community management of natural resources. This would help in proper management of energy, water and soil moisture restoration and retention and that will lead to a sustainable economic development. This is how need and perception on ecological restoration are gaining grounds among the small holdings. They are becoming more and more cost sensitive because they have to increase the return per unit area. Whenever the holding size is getting skewed and smaller, the thrust for upscaling of the income will be more. In this entire episode of energy, resource and economy management, there developed a better perception of ecological restoration.

The variable income per unit of land reflected a negative and significant relationship with the dependent variable in the study. A close inspection reveals that no matter how much behavioral change or change in practice takes place, this change has no visible and positive impact in the farmers' income from their land. This is a vital point of inspection that if return is not more or at par with the existing system, perception development alone would not be able to catalyze the uptake of new technologies like conservation agriculture.

Table 12: Path analysis of perception on ecosystem management in CA (y_4) vs. 20 exogenous variables in New Alluvial Zone of West Bengal

Variables	Total Effect (TE)	Direct Effect (DE)	Indirect Effect (IE)	Highest Indirect Individual Effect (HIIE)
Age of the respondents (x_1)	0.390	0.204	0.186	0.056(x_9)
Education (x_2)	0.190	0.025	0.165	0.052(x_9)
Family size (x_3)	-0.274	-0.168	-0.106	0.091(x_{10})
Farm size (x_4)	0.126	-0.004	0.130	0.105(x_9)
Cropping intensity (x_5)	-0.132	-0.010	-0.122	-0.068(x_9)
No. of fragments (x_6)	-0.240	-0.147	-0.093	-0.057(x_1)

Variables	Total Effect (TE)	Direct Effect (DE)	Indirect Effect (IE)	Highest Indirect Individual Effect (HIIE)
Annual income (x_7)	0.131	0.001	0.130	-0.060(x_8)
Income per capita (x_8)	0.173	-0.105	0.278	0.088(x_{10})
Income per unit of land (x_9)	-0.523	-0.279	-0.244	-0.082(x_{11})
Annual expenditure (x_{10})	-0.141	0.153	-0.294	-0.100(x_3)
Stubble height (x_{11})	0.398	0.248	0.150	0.092(x_9)
Volume of residue (x_{12})	0.291	0.130	0.161	0.076(x_9)
Scientific orientation (x_{13})	0.211	-0.007	0.218	0.068(x_9)
Innovativeness (x_{14})	0.205	0.074	0.131	0.035(x_{20})
Extension agency contact (x_{15})	0.176	0.169	0.007	0.026(x_9)
Information seeking behavior (x_{16})	0.379	0.171	0.208	0.095(x_9)
Residue management score (x_{17})	0.110	0.077	0.033	-0.076(x_{11})
Perception on natural resource degradation (x_{18})	0.297	0.098	0.199	0.046(x_{19})
No. of livestock (x_{19})	0.213	0.128	0.085	0.035(x_{18})
Mass media utilization (x_{20})	0.143	0.105	0.038	-0.042(x_{10})

Residual effect: 0.403

Table 12 presents the path analysis of dependent variable perception on ecosystem management (y_4) wherein the total effect (coefficient of correlation) has been decomposed into direct, indirect and residual effect.

The variable **income per unit of land** has exerted **highest direct effect** on the dependent variable perception on ecosystem management. **Annual expenditure** has exerted **highest indirect effect** on the dependent variable under study. **Income per unit** of land has also routed **highest indirect individual effect** on as many as 9 exogenous variable. This proves that managing ecosystem is an expensive task. Small and marginal farmers whose income is mostly restricted to the agriculture production, will hardly bother about managing ecosystem if it does not yield into better returns.

Table 13: Stepwise regression analysis of perception on ecosystem management in CA (y_4) versus 20 independent ($x_1 - x_{20}$) variables in New Alluvial Zone of West Bengal (n=75)

Variables	Reg. coeff. B	S.E. B	Beta	t value	R2	Std. error of the estimate
Income per unit of land (x_9)	0.000	0.000	-0.285	-2.901	0.498	1.444
Age (x_1)	0.040	0.014	0.250	2.849		
Stubble height (x_{11})	0.292	0.093	0.287	3.137		
Information seeking behavior (x_{16})	0.165	0.073	0.211	2.260		
Perception on natural resource degradation (x_{18})	0.030	0.014	0.194	2.192		

Table 13 shows that variables, **income per unit of land (x_9), age (x_1), stubble height retention (x_{11}), information seeking behavior (x_{16}) and perception on natural resource degradation (x_{18})** have been retained at the last step. The R^2 value being 0.498, reveals that these variables together have explained 49.8 per cent of the variance embedded in **perception on ecosystem management in CA (y_4).**

Development of perception largely depends on experience as well as information sources. Respondents having better perception on natural resource degradation tends to protect ecosystem adopting sustainable practices like maintaining stubble in the field. All this will sustain only if income is not hampered or if income is increased. Without a positive change in the income, perception development will turn futile in long run.

Canonical Covariate Analysis

The CCA have been carried out between two sets of variable x and y. Both the cross loading values and the directional characters are being considered in grouping these two sets of variables. The CCA model here depicts that the left side variable (dependent variable perception on ecosystem management (y_4)) has become rather solitary by maintaining a distance from other three dependent variables (perception on water management (Y1), perception Y2 and perception on carbon management (Y3)). While the dependent variable ecosystem management perception has developed an affinity of interactions with the right side variables (education, family size, farm size, number of fragments, income per capita, income per unit of land, scientific orientation, innovativeness, extension agency contact, residue management score and mass media utilization) to imply that in order to change or maneuver the perception on ecosystem management these variables, these variables has come up as strong determinants.

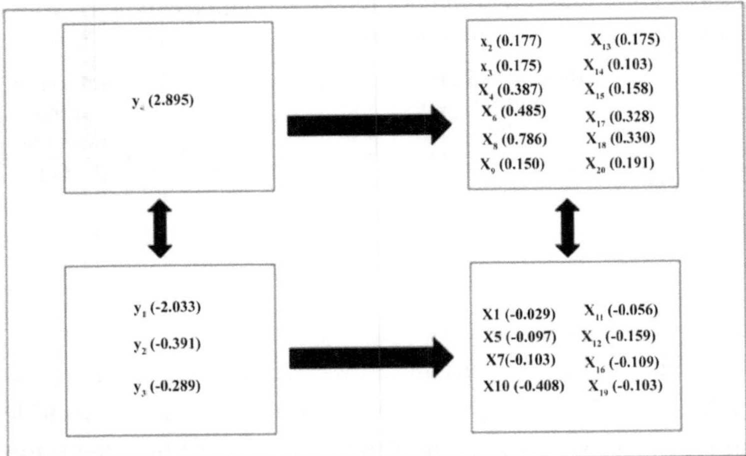

Similarly, the other three dependent variables, which have created a sub group, have recorded a clandestine and selective responses to the following variables (age of the respondent, cropping intensity, annual income, annual expenditure, stubble height retained in the field, volume of residue management, information seeking behavior and number of livestock).

CCA by becoming an approach for reciprocal regression between two sets of variables, it provides huge strategic opportunities. The 3 output variable, eg. water, nutrient and carbon management has recorded a propensity for isochronous movement when the selected right side sets of variables are supposed to create a specific impact on these 3. It is discernible that without water and nutrient management, the organic carbon level cannot be scaled up or maintained and on the other way, nutrient, organic carbon and water parameters for a soil echelon have a collective output in a response to the selected exogenous variables (right side variables).

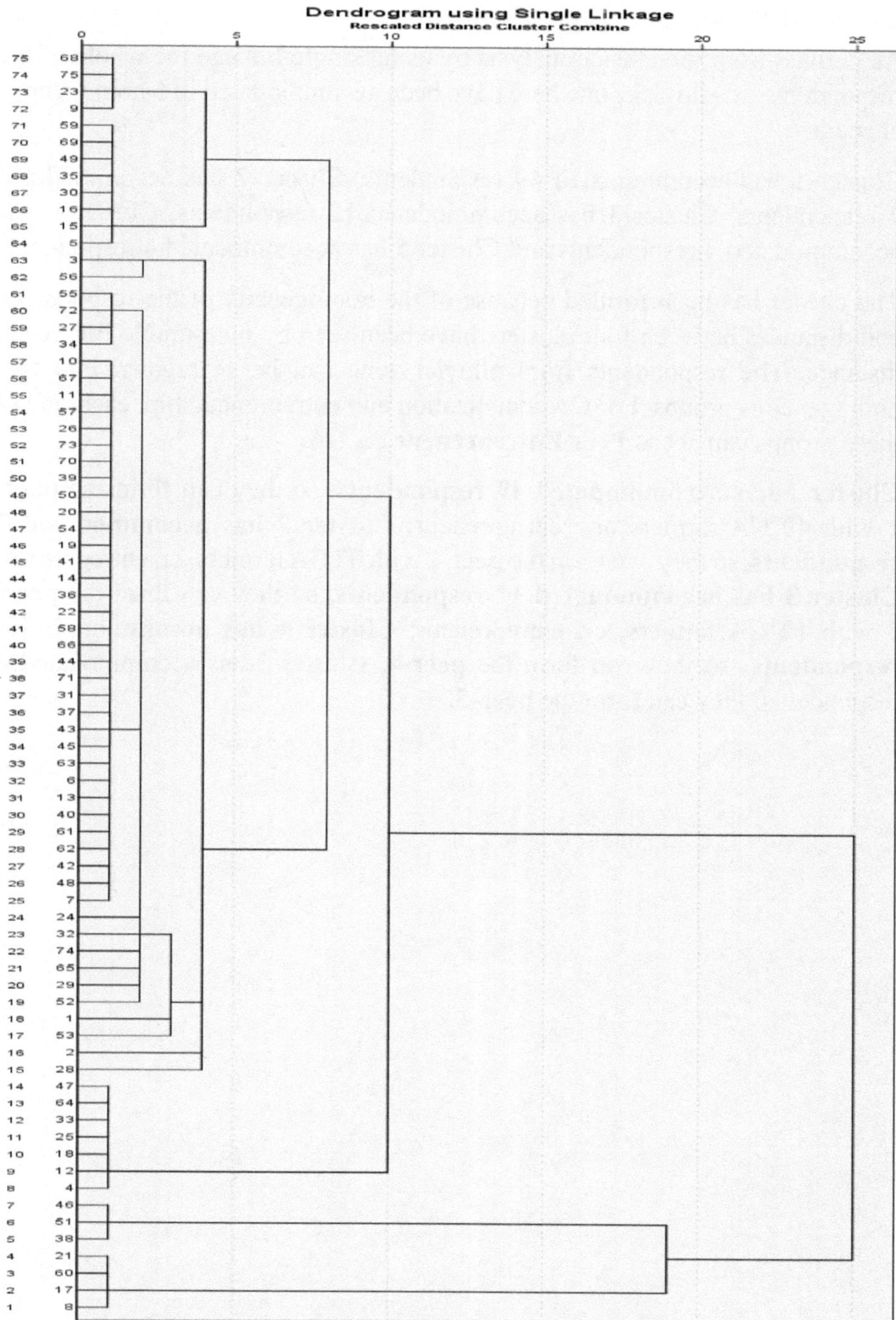

Fig. Cluster analysis for the respondents in New Alluvial Zone of West Bengal

Cluster Analysis

As derived from the cluster analysis by using single linkage the whole of the respondents of Alluvial Zone here have been accommodated in 5 homogenous clusters.

Cluster 1 has accommodated 49 respondents, Cluster 2 has accommodated 7 respondents, Cluster 3 has accommodated 12 respondents, Cluster 4 has accommodated 4 respondents and Cluster 5 has accommodated 4 respondents

The cluster has been formed because of the homogeneity of the respondents and distances between four clusters have been seen by measurable Euclidean distance. The respondents from alluvial zone can be segregated into five homogeneous groups. For CA socialization and entrepreneurship, each one of these groups can act as **Peer Entrepreneur on CA.**

Cluster 1 has accommodated 49 respondents, so they can form the **peer-1** with 49 CA farmers or , entrepreneur; **Cluster 2 has accommodated 7 respondents,** so they can form the **peer-2** with 07 CA farmers, or, entrepreneur, **Cluster 3 has accommodated 12 respondents,** so they can form the **peer-3** with 12 CA farmers, or, entrepreneur; **Cluster 4 has accommodated 4 respondents;** so they can form the **peer-4;** Cluster 5 has accommodated 4 respondents; they can form the **peer-5.**

Artificial Neural Network

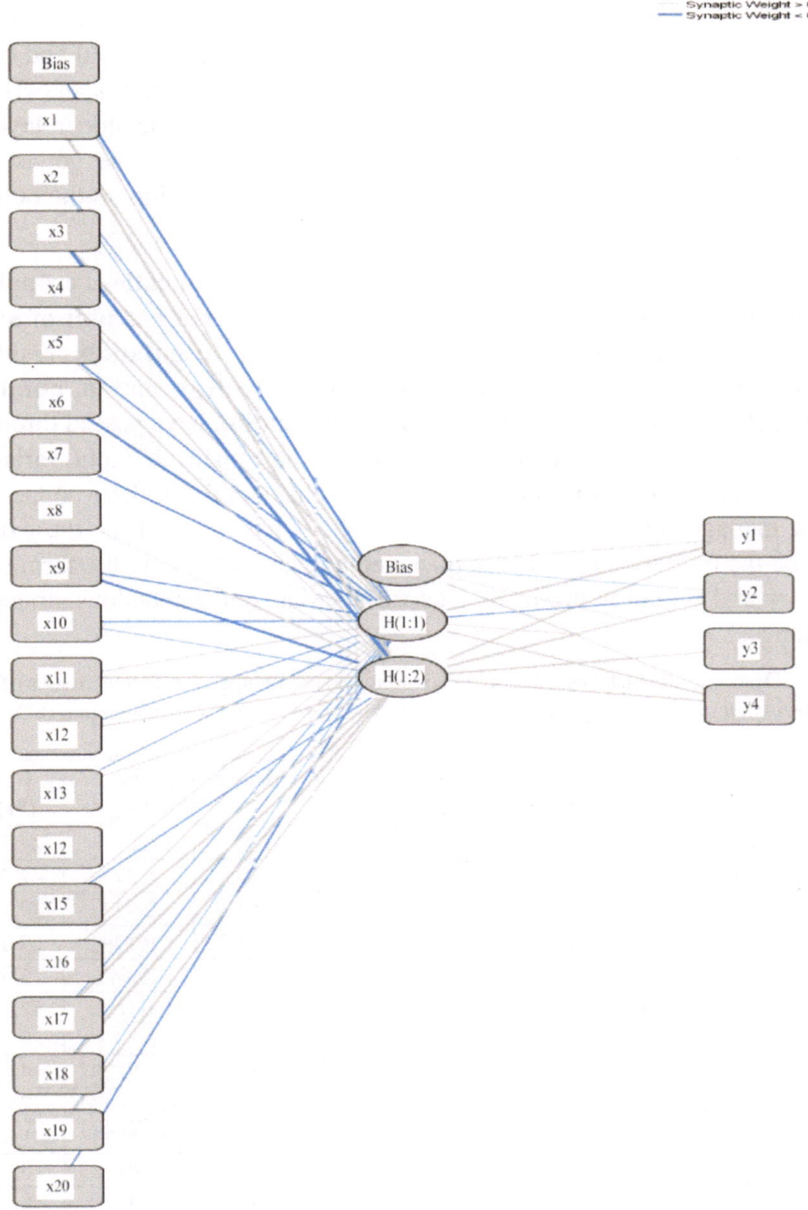

Hidden layer activatin function: Hyperbolic tangent
Output layer activation function: Identity

Fig. Artificial Neural Network of interaction on independent variables with dependent variables in New Alluvial Zone of West Bengal

ANN is the mimicry of human brain in analysing the complex network of interaction between two sets of variables viz. input variables and output variables. Out of these polyhedral interactions, some of the selected input variables have been found to generate a swashbuckling effect (bold and blue lines) on the output variables after passing through two hidden layers (H1:1 and H1:2). When these inputs variables are passing through the hidden layers, error has been rationalized to the extent possible and what we get is the substantive effect of input variables in characterizing or scaling up the output variables.

From the set of input variables the following variables **education, cropping intensity, number of fragments, annual income, income per unit of and, annual expenditure, volume of residue, scientific orientation, residue management score, perception on natural resource degradation, number of livestock owned (cattle) and mass media utilization** have passed through the hidden layer (H1:1). However, in steering the impact of input variable from left side, **number of land fragments (x6)** has got the highest impact on nutrient management. The higher is the number of fragmentation, the worse would be the nutrient and resource management, and ultimately the farm will be energy and economy prodigal. This variable has got tremendous strategic importance.

Table 14: Descriptive statistics of the socio-economic and communication characteristics of the respondents in terms of Range (Minimum and Maximum), Mean, Standard Deviation (SD) and Coefficient of Variation (CV) in Terai Zone of West Bengal

Variables	Range		Mean	SD	CV (%)
	Minimum	Maximum			
Age of the respondent (x_1)	25.00	76.00	49.43	12.227	24.74
Education (x_2)	1.00	6.00	3.30	1.377	41.76
Family size (x_3)	1.00	16.00	5.52	2.747	49.77
Farm size (x_4)	1.50	32.00	13.29	9.178	69.07
Cropping intensity (x_5)	72.00	450.00	237.29	70.447	29.69
Number of land fragments (x_6)	1.00	7.00	2.02	1.169	57.87
Annual income (x_7)	25000.00	452000.00	179635.71	80120.297	44.60
Income per capita (x_8)	5700.00	140000.00	31614.56	18064.237	57.14
Income per unit of land (x_9)	8550.00	213333.33	35138.94	28679.335	81.62
Annual expenditure (x_{10})	12000.00	240000.00	87317.14	42390.323	48.53
Stubble height (x_{11})	1.00	28.00	7.57	4.443	58.66
Volume of residue (x_{12})	325.00	26050.00	8890.64	7068.640	79.51
Scientific orientation (x_{13})	11.00	30.00	19.82	3.645	18.39
Innovativeness (x_{14})	7.00	23.00	15.41	3.490	22.65

Extension agency contact (x_{15})	5.00	16.00	10.37	2.436	23.49
Information seeking behavior (x_{16})	8.00	18.00	13.34	2.284	17.13
Residue management score (x_{17})	7.00	17.00	12.03	2.230	18.53
Perception on natural resource degradation (x_{18})	49.00	80.75	67.16	8.160	12.15
Number of livestock owned (cattle) (x_{19})	1.00	15.00	4.14	2.467	59.54
Mass media utilization (x_{20})	1.00	9.00	5.54	1.954	35.24
Perception on water management in CA (y_1)	6.00	16.00	12.25	2.118	17.29
Perception on nutrient management in CA (y_2)	8.00	16.00	13.38	1.981	14.81
Perception on carbon management in CA (y_3)	4.00	15.00	10.71	2.253	21.02
Perception on ecosystem management in CA (y_4)	6.84	15.33	12.02	1.914	15.93

Table 14 represents the distribution of variables in terms of range i.e. maximum and minimum values, mean, standard deviation and coefficient of variation for terai zone of West Bengal.

It has been found that for the independent variable **age (x_1)**, the range lies between 76 and 25 years, where the oldest respondent is of 76 years and the youngest respondent being 25 years old. The mean value of the age of the respondents in this study was found to be 49.43 years with a standard deviation of 12.22 in the dataset. Coefficient of variation for this variable was found to be 24.73 per cent, which indicates a high level of consistency in the distribution.

It has been found that for the independent variable **education (x_2)**, the range lies between the scale value 1 and 6, which means the minimum qualification amongst the respondents is illiteracy and the highest formal education received among the respondents is either a graduate or post graduate. The mean scale value lies at 3.30, which means majority of the respondents received formal education up to middle school. The standard deviation in the data set is measured at 1.37. Coefficient of variation for this variable was found to be 41.75 per cent, which indicates the level of consistency in the distribution is at higher end.

It has been found that for the independent variable **family size (x_3)**, the range stretches within a minimum of 1 and a maximum of 16 family members. The average size of family consists of approximately 6 (5.52) members.

The measured standard deviation in the dataset was found to be 2.75 and the calculated coefficient of variation is 49.77 per cent. This indicates there is a moderate to high level of consistency in distribution of data.

It has been found that for the independent variable **farm size (x_4)**, the range stretches between minimum of 1.5 biogas and maximum of 32 biogas of land. The mean value of farm size of the respondents in this study was found to be 13.29 biogas with a standard deviation of 9.18 in the dataset. Coefficient of variation for this variable was found to be 69.07 per cent, which indicates a moderate level of consistency in the distribution.

It has been found that for the independent variable **cropping intensity (x_5)**, the minimum was observed at 72.00 per cent and maximum at 450.00 per cent. The mean value of cropping intensity in this study stands at 237.29 per cent with a standard deviation of 70.45 in the dataset. Coefficient of variation observed for this variable was 29.68 per cent, which indicates a high level of consistency in the distribution.

It has been found that for the independent variable **number of land fragments (x_6)**, the range lies between 1 (minimum) and 7 (maximum). The mean value of number of land fragments observed in this study was 2.02 years with a standard deviation of 1.17. Coefficient of variation for this variable was 57.87 per cent, which indicates a moderate level of consistency in the distribution.

It has been found that for the independent variable **annual income (x_7)**, the minimum is 25000 INR and the maximum is 452000 INR. The mean value is 179635.71 INR with a standard deviation of 80120.30 in the dataset. Coefficient of variation for this variable was found to be 44.60 per cent, which indicates a medium level of consistency in the distribution.

It has been found that for the independent variable **income per capita (x_8)**, the minimum is 5700 INR and the maximum is 140000 INR. The mean value is 31614.56 INR and the standard deviation is 18064.24. Coefficient of variation for this variable was found to be 57.14 per cent, which indicates a medium level of consistency in the distribution.

For the independent variable **income per unit of land (x_9)**, it has been found the minimum is 8550.00 INR and the maximum is 213333.33 INR. The mean value is 35138.94 with a standard deviation of 28679.33 in the total distribution. Coefficient of variation for this variable was found to be 81.62 per cent, which indicates a medium level of consistency in the distribution.

It has been found that for the independent variable **annual expenditure (x_{10})**, the minimum expenditure is 12000 INR and the maximum is 240000 INR.

The mean value is 87317.14 INR with a standard deviation of 42390.32 in the total distribution. Coefficient of variation estimated at 48.55 per cent, which indicates a medium level of consistency in the distribution.

It has been found that for the independent variable **stubble height (x_{11})**, the minimum height maintained is 1 inch and the maximum is 28 inches. The mean height is 7.58 inches with a standard deviation of 4.44 in total distribution. Coefficient of variation for this variable was found to be 58.66 per cent, which indicates a moderate level of consistency in the distribution.

It has been found that for the independent variable **volume of residue (x_{12})**, the minimum amount of residue generated in a farmer's field was 325 kgs and the maximum amount was 26050 kgs. The mean amount was 8890.64 with a standard deviation of 7068.64 in the total distribution. Coefficient of variation for this variable was 79.51 per cent, which indicates a medium level of consistency in the distribution.

It has been found that for the independent variable **scientific orientation (x_{13})**, the minimum score is 11 and the maximum score obtained is 30. The mean score lies at 19.82 with a standard deviation of 3.65 in the total distribution. Coefficient of variation for this variable was found to be 18.39 per cent, which indicates a very high level of consistency in the distribution.

It has been found that for the independent variable **innovativeness (x_{14})**, the minimum score is 7 and the maximum score obtained is 23. The mean score lies at 15.41 with a standard deviation of 3.49 in the total distribution. Coefficient of variation for this variable was found to be 22.65 per cent, which indicates a high level of consistency in the distribution.

It has been found that for the independent variable **extension agency contact (x_{15})**, the minimum score is 5 and the maximum score obtained is 16. The mean score lies at 10.37 with a standard deviation of 2.44 in the total distribution. Coefficient of variation for this variable was found to be 23.49 per cent, which indicates a high level of consistency in the distribution.

It has been found that for the independent variable **information seeking behaviour (x_{16})**, the minimum score is 8 and the maximum score obtained is 18. The mean score lies at 13.34 with a standard deviation of 2.28 in the total distribution. Coefficient of variation for this variable was found to be 17.13 per cent, which indicates a very high level of consistency in the distribution.

It has been found that for the independent variable **residue management score (x_{17})**, the minimum score is 7 and the maximum score obtained is 17. The mean score lies at 12.03 with a standard deviation of 2.23 in the total distribution.

Coefficient of variation for this variable was found to be 18.53 per cent, which indicates a very high level of consistency in the distribution.

It has been found that for the independent variable **perception on natural resource degradation (x18)**, the minimum score is 49.00 and the maximum score obtained is 80.75. The mean score lies at 67.16 with a standard deviation of 8.16 in the total distribution. Coefficient of variation for this variable was found to be 12.15 per cent, which indicates a high level of consistency in the distribution.

It has been found that for the independent variable **number of livestock owned (x_{19})**, the minimum count is 1 and the maximum is 15. The mean count is approximately 4 (4.14) with a standard deviation of 2.47 in the total distribution. Coefficient of variation for this variable was found to be 59.54 per cent, which indicates a medium level of consistency in the distribution.

It has been found that for the independent variable **mass media utilization (x_{20})**, the minimum score is 1 and the maximum score obtained is 9. The mean score lies at 5.55 with a standard deviation of 1.95 in the total distribution. Coefficient of variation for this variable was found to be 35.24 per cent, which indicates a high level of consistency in the distribution.

It has been found that for the dependent variable **perception on water management in CA (y_1)**, the minimum score is 6 and the maximum score obtained is 16. The mean score lies at 12.25 with a standard deviation of 2.12 in the total distribution. Coefficient of variation for this variable was found to be 17.29 per cent, which indicates a very high level of consistency in the distribution.

It has been found that for the dependent variable **perception on nutrient management in CA (y_2)**, the minimum score is 8 and the maximum score obtained is 16. The mean score lies at 13.38 with a standard deviation of 1.98 in the total distribution. Coefficient of variation for this variable was found to be 14.81 per cent, which indicates a very high level of consistency in the distribution.

It has been found that for the dependent variable **perception on carbon management in CA (y_3)**, the minimum score is 4 and the maximum score obtained is 15. The mean score lies at 10.71 with a standard deviation of 2.25 in the total distribution. Coefficient of variation for this variable was found to be 21.02 per cent, which indicates a high level of consistency in the distribution.

It has been found that for the dependent variable **perception on ecosystem management in CA (y_4)**, the minimum score is 6.84 and the maximum score obtained is 15.33. The mean score lies at 12.02 with a standard deviation of 1.91 in the total distribution. Coefficient of variation for this variable was found to be 15.93 per cent, which indicates a very high level of consistency in the distribution

Table 15: Correlation Coefficient of perception on water management (y_1) vs. 20 independent variables (x_1-x_{20}) (n=154)

Independent Variables	**'r' Value**	**Remarks**
Age of the respondent (x_1)	-0.0004	
Education (x_2)	0.266	**
Family size (x_3)	-0.052	
Farm size (x_4)	0.132	
Cropping intensity (x_5)	0.413	**
Number of land fragments (x_6)	0.103	
Annual income (x_7)	0.139	
Income per capita (x_8)	-0.269	**
Income per unit of land (x_9)	0.008	
Annual expenditure (x_{10})	0.099	
Stubble height (x_{11})	0.435	**
Volume of residue (x_{12})	0.069	
Scientific orientation (x_{13})	0.449	**
Innovativeness (x_{14})	0.472	**
Extension agency contact (x_{15})	0.161	*
Information seeking behavior (x_{16})	0.218	**
Residue management score (x_{17})	0.027	
Perception on natural resource degradation (x_{18})	0.025	
Number of livestock owned (cattle) (x_{19})	-0.119	
Mass media utilization (x_{20})	0.297	**

**Correlation is significant at the 0.01 level
*Correlation is significant at the 0.05 level

Table 15: presents the correlation coefficient of correlation between perception on water management (y_1) and 20 independent variables (x_1-x_{20}). It is discernable from the table that from the variables, education, cropping intensity, stubble height, scientific orientation, innovativeness, extension agency contact, information seeking behavior and mass media utilization have recorded positive and significant correlation with the dependent variable under discussion. The variable income per capita recorded a negative but significant correlation with the same dependent variable under study.

For the last 10 years CA has been in practice there in terai zone. The cosmopolite behavior and education of the farmers has got an edge in orienting the farmers scientifically towards management of water resources. This findings is supported by the revelations made by **Ntshangase et al. (2018)** confirms that education plays a vital role in perception development that ultimately culminate into adoption of no-till practices. The well-designed extension approaches of state agriculture university (SAU) in collaboration with national and international organizations like CIMMYT and local Farmers' Producers Organization (FPOs) has got a remarkable impact in perception development of the farmers on water management through conservation agriculture. The negative correlation between income per capita and perception on water management depicts that, if changes in the income is not reflected, farmers would hardly go for managing resources like water at any extra cost.

Table 16: Path analysis of perception on water management in CA (y_1) vs. 20 exogenous variables in Terai zone of West Bengal

Exogenous Variables	Total Effect	Direct Effect	Indirect Effect	Highest Indirect Effect
Age of the respondent (x_1)	0.0004	0.176	-0.176	-0.043(x_{13})
Education (x_2)	0.266	0.171	0.095	-0.031(x_1)
Family size (x_3)	-0.052	-0.237	0.185	0.079(x_8)
Farm size (x_4)	0.132	0.000	0.132	0.047(x_2)
Cropping intensity (x_5)	0.413	0.205	0.208	0.069(x_{13})
Number of land fragments (x_6)	0.103	0.109	-0.006	-0.071(x_3)
Annual income (x_7)	0.139	0.141	-0.002	-0.081(x_3)
Income per capita (x_8)	-0.269	-0.238	-0.031	0.079(x_3)
Income per unit of land (x_9)	0.008	-0.050	0.058	-0.030(x_6)
Annual expenditure (x_{10})	0.099	0.044	0.055	-0.104(x_3)
Stubble height (x_{11})	0.435	0.249	0.186	0.044(x_{13})
Volume of residue (x_{12})	0.069	-0.116	0.185	0.050(x_{13})
Scientific orientation (x_{13})	0.449	0.243	0.206	0.058(x_5)
Innovativeness (x_{14})	0.472	0.201	0.271	0.080(x_8)
Extension agency contact (x_{15})	0.161	0.132	0.029	0.038(x_{13})
Information seeking behavior (x_{16})	0.218	0.067	0.151	0.047(x_5)
Residue management score (x_{17})	0.027	-0.049	0.076	0.028(x_{13})
Perception on natural resource degradation (x_{18})	0.025	-0.065	0.090	-0.034(x_{12})
Number of livestock (x_{19})	-0.119	-0.078	-0.041	-0.033(x_{11})
Mass media utilization (x_{20})	0.297	0.022	0.275	0.075(x_{13})

Residual effect: 0.405

Table 16 presents the path analysis of dependent variable perception on water management (y_1) wherein the total effect (coefficient of correlation) has been decomposed into direct, indirect and residual effect.

Stubble height has exerted the highest direct effect on the dependent variable perception on water management (y_1). This is interesting to note that higher perception on water management is being contributed by another agro-ecological behavior of farmers that is maintain stubble height in their field. This is especially true for *aman* rice which is specially grown in water stagnant condition. That is how stubble height is being determined by the water regime in the rice field during harvesting.

Highest indirect effect has been recorded by mass media exposure which indicates that with higher mass media exposure, the better water perception has been configured. Scientific orientation when gets inculcated with the farmers' psyche, the ecological management and behavior of farms keeps scaling up.

Table 17: Stepwise regression analysis of perception on water management in CA (y_1) versus 20 independent ($x_1 - x_{20}$) variables in terai zone of West Bengal (n=154)

Variables	Reg. coeff. B	S.E. B	Beta	t value	R^2	Std. error of estimate
Innovativeness (x_{14})	0.188	0.037	0.309	5.112	0.516	1.509
Scientific orientation (x_{13})	0.134	0.036	0.231	3.706		
Stubble height (x_{11})	0.131	0.029	0.273	4.540		
Cropping intensity (x_5)	0.006	0.002	0.213	3.486		
Education (x_2)	0.226	0.090	0.146	2.506		
Extension agency contact (x_{15})	0 113	0.051	0.129	2.209		

Table 17 shows that variables, **innovativeness, scientific orientation, stubble height, cropping intensity, education** and **extension agency contact** have been retained at the last step. The R^2 value being 0.516, reveals that these three variables together explains 51.6 per cent of the variance embedded in **perception on water management in CA (y_1)** in terai region of West Bengal.

To learn advanced and sustainable farming technology, education is a key element. Proper knowledge and adequate support from the extension agents helps to in developing scientific outlook and innovativeness in the farmers. This changed outlook would help in perceptional change of the farmers towards water management in their respective farms. Also cropping intensity is a determining factor in perceptional development in farmers as those who farm intensively will make more efficient use of resources in comparison to those who till once or twice a year.

Table 18: Correlation Coefficient of perception on nutrient management (y_2) vs. 20 independent variables (x_1-x_{20}) in Terai Zone of West Bengal (n=154)

Independent Variables	'r' Value	Remarks
Age of the respondent (x_1)	0.378	**
Education (x_2)	0.138	
Family size (x_3)	-0.048	
Farm size (x_4)	0.460	**
Cropping intensity (x_5)	-0.038	
Number of land fragments (x_6)	0.025	
Annual income (x_7)	0.142	
Income per capita (x_8)	0.042	
Income per unit of land (x_9)	0.094	
Annual expenditure (x_{10})	0.077	
Stubble height (x_{11})	0.108	
Volume of residue (x_{12})	0.476	**
Scientific orientation (x_{13})	0.030	
Innovativeness (x_{14})	0.040	
Extension agency contact (x_{15})	0.045	
Information seeking behavior (x_{16})	0.096	
Residue management score (x_{17})	0.313	**
Perception on natural resource degradation (x_{18})	0.415	**
Number of livestock (x_{19})	0.009	
Mass media utilization (x_{20})	0.056	

**Correlation is significant at the 0.01 level

*Correlation is significant at the 0.05 level

Table 18 presents the correlation coefficient of correlation between perception on nutrient management (y_2) and 20 independent variables (x_1-x_{20}). It is discernable from the table that from the variables, age, farm size, volume of residue, residue management score and perception on natural resource degradation have recorded positive and significant correlation with the dependent variable, perception on nutrient management under discussion.

On farm nutrient management is an important sustainable practice. Using farm residues as nutrient sources requires experiences that comes with age of practice. More the residue generated on farm, better the management of the residues, more the knowledge on natural resource degradation, it is quite natural to develop a better perception on nutrient management among the farmers. Bigger farm size helps in more residue generation, which when managed properly helps in management of nutrient application as well.

Table 19: Path analysis of perception on nutrient management in CA (y_2) vs. 20 exogenous variables in Terai Zone of West Bengal

Exogenous Variables	Total Effect (TE)	Direct Effect (DE)	Indirect Effect (IE)	Highest Indirect Individual Effect (HIIE)
Age of the respondent (x_1)	0.378	0.274	0.104	0.058(x_{12})
Education (x_2)	0.138	0.062	0.076	0.082(x_4)
Family size (x_3)	-0.048	-0.142	0.094	0.056(x_4)
Farm size (x_4)	0.460	0.298	0.162	0.115(x_{12})
Cropping intensity (x_5)	-0.038	-0.074	0.036	-0.041(x_1)
Number of land fragments (x_6)	0.025	-0.077	0.102	0.069(x_{12})
Annual income (x_7)	0.142	-0.023	0.165	0.091(x_4)
Income per capita (x_8)	0.042	0.026	0.016	0.047(x_3)
Income per unit of land (x_9)	0.094	0.048	0.046	-0.031(x_{12})
Annual expenditure (x_{10})	0.077	0.049	0.028	0.077(x_{12})
Stubble height (x_{11})	0.108	0.055	0.053	0.031(x_4)
Volume of residue (x_{12})	0.476	0.305	0.171	0.113(x_4)
Scientific orientation (x_{13})	0.030	-0.043	0.073	0.063(x_{12})
Innovativeness (x_{14})	0.040	-0.033	0.073	0.037(x_4)
Extension agency contact (x_{15})	0.045	-0.060	0.105	0.075(x_{12})
Information seeking behavior (x_{16})	0.096	0.050	0.046	0.018(x_4)
Residue management score (x_{17})	0.313	0.154	0.159	0.054(x_4)
Perception on natural resource degradation (x_{18})	0.415	0.170	0.245	0.093(x_4)
Number of livestock (x_{19})	0.009	-0.109	0.118	0.048(x_4)
Mass media utilization (x_{20})	0.056	0.033	0.023	0.030(x_4)

Residual effect: 0.466

Table 19 presents the path analysis of dependent variable perception on water management (y_1) wherein the total effect (coefficient of correlation) has been decomposed into direct, indirect and residual effect.

The variable volume of residue has exerted highest direct effect on perception on nutrient management. Nutrient management and on farm crop residue management are interlinked since the higher volume of crop residues, the nutrient status in the soil regime keeps improving. When residues are well decomposed, it keeps contributing to OC status, which in turn increases the microbial count and it makes nutrients available at the root zone.

Perception on natural resource degradation has exerted highest indirect effect on perception on nutrient management. Perception on natural resource degradation creates a general concern amongst the farmers for this zone and it makes the farmer more committed towards resource monitoring and management in their respective farms.

Better nutrient management has been reflected in bigger size farms. That's why it has been found that the variable farm size has routed the highest indirect effect on as many as 10 exogenous variable to characterize the consequent variable, perception on nutrient management. Farm size reflects resources capability and enterprise opportunity. These two in turn make the farmers more caring and attentive for nutrient management.

Residual effect being 0.466 means that 46.6% of the variance couldn't been explained with these combination of 20 exogenous variables.

Table 20: Stepwise regression analysis of perception on nutrient management in CA (y_2) versus 20 independent ($x_1 - x_{20}$) variables in Terai Zone of West Bengal (n=154)

Variables	Reg. coeff. B	S.E. B	Beta	t value	R^2	Std. error of the estimate
Volume of residue (x_{12})	0.000	0.000	0.253	3.839	0.492	1.445
Farm size (x_4)	0.065	0.014	0.298	4.472		
Age (x_1)	0.045	0.010	0.278	4.573		
Perception on natural resource degradation (x_{18})	0.045	0.016	0.184	2.873		
Family size (x_3)	-0.116	0.044	-0.160	-2.659		
Residue management score (x_{17})	0.129	0.054	0.144	2.370		

Table 20 shows that variables, **volume of residue, farm size, age, perception on natural resource degradation, family size and residue management score** have been retained at the last step. The R^2 value being 0.492, reveals that these two variables together explains 49.2 per cent of the variance embedded in **perception on nutrient management in CA (y_2)** in terai region of West Bengal.

Perception on natural resource degradation has got a direct impact on managing natural resources like residues for managing soil nutrition. With bigger farms and more residues, this gives ample scope to farmers to manage nutrients at farm recycling of the residues.

Table 21: Correlation Coefficient of perception on carbon management (y_3) vs. 20 independent variables (x_1-x_{20}) in Terai Zone of West Bengal (n=154)

Independent Variables	'r' Value	Remarks
Age of the respondent (x_1)	-0.067	
Education (x_2)	0.391	**
Family size (x_3)	-0.054	
Farm size (x_4)	0.140	
Cropping intensity (x_5)	0.289	**
Number of land fragments (x_6)	0.044	

Independent Variables	'r' Value	Remarks
Annual income (x_7)	0.244	**
Income per capita (x_8)	-0.096	
Income per unit of land (x_9)	0.129	
Annual expenditure (x_{10})	0.130	
Stubble height (x_{11})	0.352	**
Volume of residue (x_{12})	0.090	
Scientific orientation (x_{13})	0.363	**
Innovativeness (x_{14})	0.185	*
Extension agency contact (x_{15})	0.237	**
Information seeking behavior (x_{16})	0.049	
Residue management score (x_{17})	0.086	
Perception on natural resource degradation (x_{18})	0.010	
Number of livestock (x_{19})	-0.123	
Mass media utilization (x_{20})	0.499	**

**Correlation is significant at the 0.01 level
*Correlation is significant at the 0.05 level

Table 21 presents the correlation coefficient of correlation between perception on carbon management (y_3) and 20 independent variables (x_1-x_{20}). It is discernable from the table that from the variables, education, cropping intensity, annual income, stubble height, scientific orientation, innovativeness, extension agency contact and mass media utilization have recorded positive and significant correlation with the dependent variable, perception on carbon management (y_3) under discussion.

Carbon management is a complex yet important thing that needs to be taken care of in maintaining the sustainability of the land in long run. To build up an effective perception on carbon management one needs to have education to develop a scientific temperament. This scientific temperament is often developed through exposure to different cosmopolite sources of information including mass media and contact with extension agents. Innovative nature of farmers often test their knowledge through experimentation. Normally high cropping intensity predicts better returns. Therefore if returns is better, motivated farmers would involve in managing organic carbon in their fields realizing its contribution in assuring better returns. This finding is line with other research finding which says information is an important predictor is adoption of new agricultural technologies **(Jabbar et al., 2003).**

Table 22: Path analysis of perception on carbon management in CA (y_3) vs. 20 exogenous variables($x_1 - x_{20}$) in Terai Zone of West Bengal (n=154)

Exogenous Variables	**Total Effect**	**Direct Effect**	**Indirect Effect**	**Highest Indirect Effect**
Age of the respondent (x_1)	-0.067	0.092	-0.159	-0.059(x_2)
Education (x_2)	0.391	**0.338**	0.053	**0.029(x_{20})**
Family size (x_3)	-0.054	-0.167	0.113	0.074(x_7)
Farm size (x_4)	0.140	-0.027	0.167	0.093(x_2)
Cropping intensity (x_5)	0.289	0.160	0.129	**0.039(x_{20})**
Number of land fragments (x_6)	0.044	0.033	0.011	0.066(x_7)
Annual income (x_7)	0.244	0.216	0.028	-0.057(x_3)
Income per capita (x_8)	-0.096	-0.122	0.026	0.057(x_7)
Income per unit of land (x_9)	0.129	0.025	0.104	**0.039(x_{20})**
Annual expenditure (x_{10})	0.130	-0.004	0.134	0.152(x_7)
Stubble height (x_{11})	0.352	0.170	0.182	**0.067(x_{20})**
Volume of residue (x_{12})	0.090	-0.047	0.137	0.062(x_7)
Scientific orientation (x_{13})	0.363	0.105	**0.258**	**0.091(x_{20})**
Innovativeness (x_{14})	0.185	-0.045	0.229	**0.045(x_{20})**
Extension agency contact (x_{15})	0.237	0.148	0.089	**0.078(x_{20})**
Information seeking behavior (x_{16})	0.049	-0.054	0.103	0.037(x_5)
Residue management score (x_{17})	0.086	-0.005	0.091	0.037(x_2)
Perception on natural resource degradation (x_{18})	0.010	-0.044	0.054	0.049(x_2)
Number of livestock (x_{19})	-0.123	-0.162	0.039	0.053(x_2)
Mass media utilization (x_{20})	0.499	0.295	0.204	0.039(x_{15})

Residual effect: 0.470

Table 22 presents the path analysis of dependent variable perception on carbon management (y_3) wherein the total effect (coefficient of correlation) has been decomposed into direct, indirect and residual effect.

The variable **education** has exerted **highest direct effect** on perception on carbon management. Perception on organic carbon management needs educational inputs in the mindset of farmers. The terai zone is classically deficient of OC due its core of biophysical and chemical characters of its soil ecology. That's why young enterprising farmers in North Bengal are getting more involved for carbon management through addition of crop residues and left out biomasses following the harvest of a crop.

The variable **scientific orientation** has exerted **highest indirect effect** on perception on carbon management (y_3). Modern agricultural management needs a fair amount of scientific orientation from the participating farmers. The higher is the scientific orientation of the farmers, it has been found the better has been the resource conservation, management and monitoring. That is why this variable has routed highest indirect effect.

It has been found that variable **mass media utilization** has routed **highest indirect individual effect** on as many as **seven exogenous variables.**

The residual effect being 0.470 means that 47% of the variance couldn't been explained with these combination of 20 exogenous variables.

Table 23: Stepwise regression analysis of perception on carbon management in CA (y_3) versus 20 independent ($x_1 - x_{20}$) variables in Terai Zone of West Bengal (n=154)

Variables	Reg. coeff. B	S.E. B	Beta	t value	R^2	Std. error of the estimate
Mass media utilization (x_{20})	0.381	0.076	0.329	5.039	0.489	1.653
Education (x_2)	0.541	0.100	0.330	5.404		
Stubble height (x_{11})	0.089	0.032	0.175	2.761		
Scientific orientation (x_{13})	0.072	0.041	0.117	1.785		
Extension agency contact (x_{15})	0.130	0.058	0.140	2.243		
No. of livestock owned (cattle) (x_{19})	-0.136	0.056	-0.149	-2.417		
Cropping intensity (x_5)	0.005	0.002	0.142	2.250		

Table 23 shows that variables, **mass media utilization, education, stubble height, scientific orientation, extension agency contact, no. of livestock and cropping intensity** have been retained at the last step. The R^2 value being 0.489, reveals that these retained variables together explains 48.9 per cent of the variance embedded in **perception on carbon management in CA (y_3)** in terai region of West Bengal.

Stubble height is directly related with volume of residue generation on field. Volume of residue has got a positive impact on carbon score of the land. Thus retention of stubble height is justly a major contributor to the perception on carbon management. On the other hand, better education develops better understanding on ecology and ecological resilience. Individual who understands ecology and its importance, would definitely have a better perception on carbon management and its impact on the land. Innovativeness of a farmer leads to experimentation with different crop combination not only to increase his income from land but also to improve its bio-physical condition to sustain the production in the future. Thus more innovative a farmer is, better is the cropping intensity and more is the income. It implies that whenever one builds up a better understanding on carbon management, its need and relevance, the ancillary changes in the positive direction would also be forthcoming, like yield, ecological resilience, improved income and ultimately the most expected sustainability.

Table 24: Correlation Coefficient of perception on ecosystem management in CA (y_4) vs. 20 independent variables (x_1-x_{20}) in Terai Zone of West Bengal (n=154)

Independent Variables	'r' Value	Remarks
Age of the respondent (x_1)	-0.004	
Education (x_2)	0.347	**
Family size (x_3)	-0.078	
Farm size (x_4)	0.129	
Cropping intensity (x_5)	0.366	**
Number of land fragments (x_6)	0.073	
Annual income (x_7)	0.184	*
Income per capita (x_8)	-0.197	*
Income per unit of land (x_9)	0.049	
Annual expenditure (x_{10})	0.108	
Stubble height (x_{11})	0.404	**
Volume of residue (x_{12})	0.082	
Scientific orientation (x_{13})	0.444	**
Innovativeness (x_{14})	0.368	**
Extension agency contact (x_{15})	0.196	*
Information seeking behavior (x_{16})	0.122	
Residue management score (x_{17})	0.063	
Perception on natural resource degradation (x_{18})	0.018	
Number of livestock (x_{19})	-0.155	
Mass media utilization (x_{20})	0.432	**

**Correlation is significant at the 0.01 level

*Correlation is significant at the 0.05 level

Table 24 presents the coefficient of correlation between perception on ecosystem management in conservation agriculture (y_4) and 20 independent (x_1 - x_{20}) variables. It has been found that variables viz., education, cropping intensity, annual income, stubble height, scientific orientation, innovativeness, extension agency contact and mass media utilization have recorded positive and significant correlation with the dependent variable under study. The variable income per capita showed a negative but significant correlation with the dependent variable under study.

Education helps to build up proper understanding on ecological surrounding and function. The new generation, young farmers have already been exposed, in their school pedagogy or training experiences on organic farming vis-à-vis CA, are more prone to protecting ecological aspects of farming. This offers a silver line for the success of CA.

Higher level of cropping intensity and inclusion of eco-friendly crops in the intercropping practices, the prospect of ecological management including soil, water and biodiversity are gaining stronger grounds. High intensity farming, after attaining highest possible productivity per unit area, shows a natural propensity to go after CA to restore whatever has already been depleted to support high yield behavior of farming.

Since terai agriculture has already shown an increasing propensity towards CA, as reflected faster socialization of CA, it is good to see that higher socialization of ecological management practices has been supported by increased income. The products of CA have also found market pathways as well as network, which has already been defended by micro level organizations like FPO in this part of WB.

Whenever Conservation Agriculture has started gaining ground, it is obvious its different integral component will also be positively impacted. That is how, terai agriculture has started showing elements and ingredients of conservation agriculture in the form of admixing crop residues, retention of stubbles, practices of mulching and as a whole a neo belligerent for socializing conservation agriculture and as a delight to see it has been happening along the topography of farm economy.

Table 25: Path analysis of perception on ecosystem management in CA (y_4) vs. 20 exogenous variables ($x_1 - x_{20}$) variables in Terai Zone of West Bengal (n=154)

Exogenous Variables	Total Effect (TE)	Direct Effect (DE)	Indirect Effect (IE)	Highest Indirect Individual Effect (HIIE)
Age of the respondent (x_1)	-0.004	0.186	-0.190	-0.049(x_2)
Education (x_2)	0.347	0.280	0.067	-0.033(x_1)
Family size (x_3)	-0.078	-0.243	0.165	0.061(x_8)
Farm size (x_4)	0.129	-0.025	0.154	0.077(x_2)
Cropping intensity (x_5)	0.366	0.190	0.176	0.058(x_{13})
Number of land fragments (x_6)	0.073	0.067	0.006	-0.072(x_3)
Annual income (x_7)	0.184	0.173	0.011	-0.083(x_3)
Income per capita (x_8)	-0.197	-0.184	-0.013	0.081(x_3)
Income per unit of land (x_9)	0.049	-0.051	0.100	0.031(x_7)
Annual expenditure (x_{10})	0.108	0.047	0.061	0.122(x_7)
Stubble height (x_{11})	0.404	0.199	0.205	0.042(x_{20})
Volume of residue (x_{12})	0.082	-0.093	0.175	0.050(x_7)
Scientific orientation (x_{13})	0.444	0.207	0.237	0.057(x_{20})
Innovativeness (x_{14})	0.368	0.104	0.264	0.062(x_8)
Extension agency contact (x_{15})	0.196	0.134	0.062	0.049(x_{20})
Information seeking behavior (x_{16})	0.122	-0.015	0.137	0.044(x_5)
Residue management score (x_{17})	0.063	-0.024	0.087	0.031(x_2)
Perception on natural resource degradation (x_{18})	0.018	-0.056	0.074	0.040(x_2)
Number of livestock (x_{19})	-0.155	-0.154	-0.001	0.044(x_2)
Mass media utilization (x_{20})	0.432	0.186	0.246	0.064(x_{13})

Residual effect: 0.413

Table 25 presents the path analysis of dependent variable perception on ecosystem management (y_4) wherein the total effect (coefficient of correlation) has been decomposed into direct, indirect and residual effect.

Exogenous variable **education** has exerted **highest direct effect** on perception on ecosystem management. Education promotes acquisition of skills and here in this case, skill in managing an ecosystem by a farmer. All the three basic functions, provisioning services (any type of benefit that can be extracted from nature, eg. Food, timber, etc), regulatory (ecosystem provide many of the basic services that makes plausible for environment eg., bees pollinate flowers, tree roots hold soil to erosion) and cultural services (importance of ecosystem in human mind can be traced back to ancient times.) supporting services (photosynthesis, creation of needs proper education. That's how it has been found that education has exerted the highest direct effect on ecosystem management.

Innovativeness has exerted **highest indirect effect** on perception on ecosystem management (y_4). Ecosystem management by becoming complex and diverse needs seamless flow of innovations. Into its mentoring and regeneration. That is how highest indirect effect has been exerted by innovativeness.

The variable **education** has routed **highest indirect individual effect** in as many as 5 exogenous variable.

The residual effect being 0.413 means that 41.3% of the variance couldn't been explained with these combination of 20 exogenous variables.

Table 26: Stepwise regression analysis of perception on ecosystem management in CA (y_4) versus 20 independent ($x_1 - x_{20}$) variables in Terai Zone of West Bengal (n=154)

Variables	Reg. coeff. B	S.E. B	Beta	t value	R^2	Std. error of the estimate
Scientific orientation (x_{13})	0.117	0.034	0.221	3.461	0.527	1.357
Stubble height (x_{11})	0.079	0.026	0.184	3.003		
Education (x_2)	0.391	0.084	0.280	4.663		
Mass media utilization (x_{20})	0.241	0.060	0.245	4.005		
Innovativeness (x_{14})	0.082	0.034	0.148	2.426		
Cropping intensity (x_5)	0.006	0.002	0.207	3.354		
No. of livestock owned (cattle) (x_{19})	-0.108	0.047	-0.138	-2.296		
Age (x_1)	0.021	0.009	0.134	2.250		

Table 26 shows that variables, **scientific orientation, stubble, education, mass media utilization, innovativeness, cropping intensity, number of livestock owned and age** have been retained at the last step. The R^2 value being 0.527, reveals that these three variables together explains 52.7 per cent of the variance embedded in **perception on ecosystem management in CA (y_4)** in terai region of West Bengal.

Ecosystem management involves management of water, soil and biodiversity as well as eco-friendly entrepreneurship. All these demands a good scientific orientation of the practicing farmer. Education and exposure to mass media helps to develop the thought process generating into perception development. Retention of the variables confirms their substantive contribution in developing the perception on ecosystem management of the farmer.

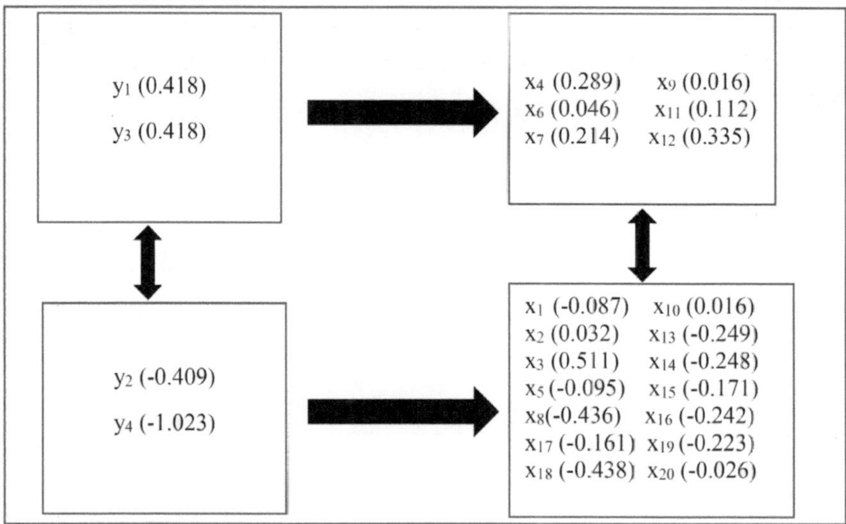

Fig. Canonical Covariate Analysis for the respondents in Terai Zone of West Bengal

Canonical Covariate Analysis for respondents of Terai Zone of West Bengal

CCA is another extension of multiple regression where rather than using a single outcome variable Y, two or more Y variables are predicted by two or more predictor X variables. The purpose is usually not to predict Ys from Xs but to explain the relationship between the X and Y variable sets. The analysis can be run in both directions, Xs predicting Ys and Ys predicting Xs, but the researcher is generally concerned about Xs predicting Ys where Y is the new hypothesized inter-relationship and the X set are the traditional predictors seen in prior research.

The CCA (terai) model has been executed for the terai zone shows that there has been a sub conglomeration of two dependent variable perception on water management (y_1) and perception on carbon management (y_3). It may so happened that there might be a congenital bondage between perception on water management and perception on carbon management and when these are moving together, following variables from the right side variables set **farm size, number of fragments, annual income, income per unit of land, stubble height and volume of residue** are precisely influencing the depending variables. Similarly, for the left side dependent variables perception on nutrient management (y_2) and perception on ecosystem management (y_4) are moving together and when they are forming a homogenous agglomeration with the right side variables **age, education, family size, cropping intensity, income per capita, annual expenditure, scientific orientation, innovativeness, extension agency contact, information seeking behavior, number of livestock and mass media utilization** are precisely influencing the dependent variables.

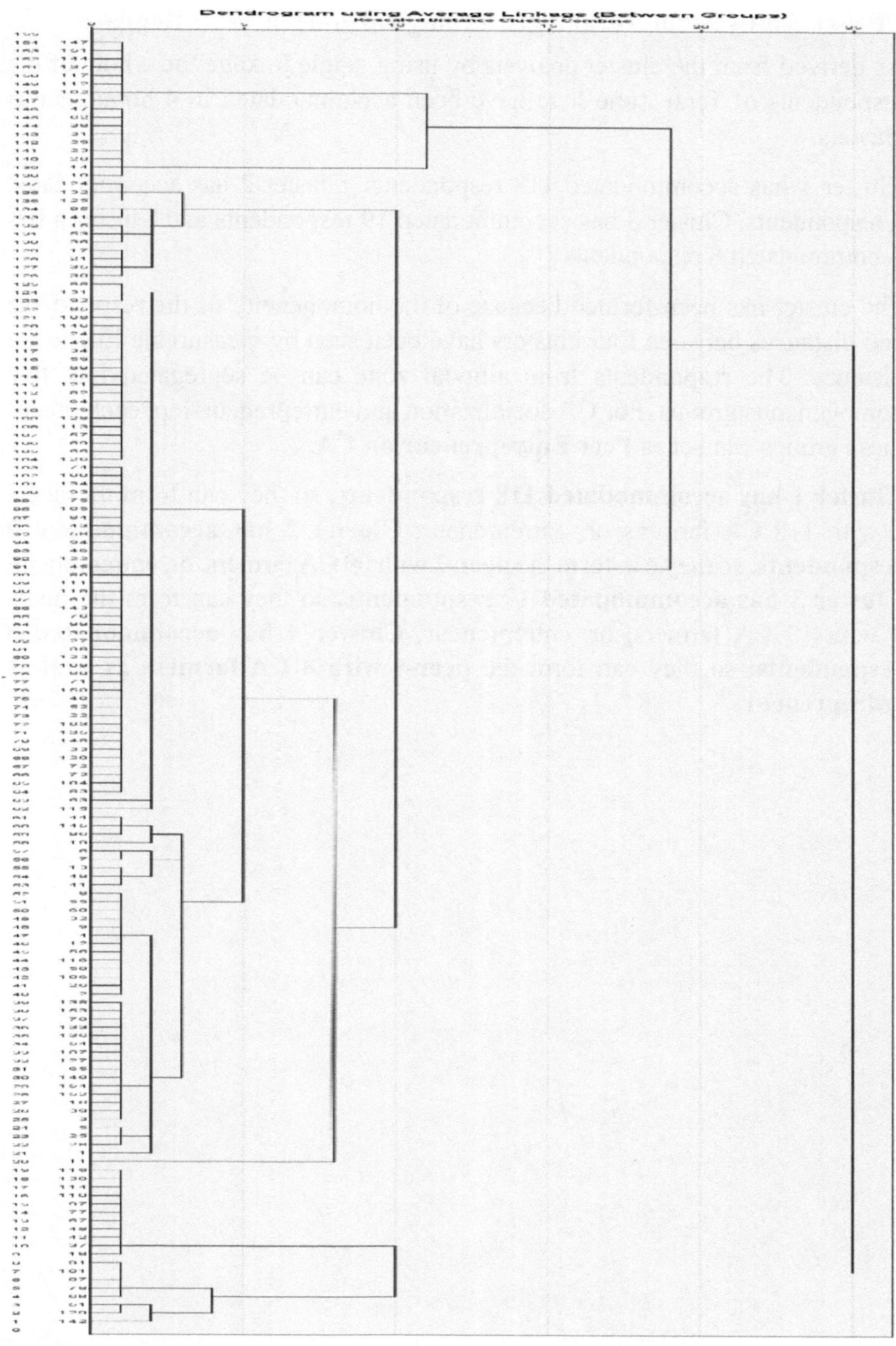

Fig. Cluster analysis for the respondents in Terai Zone of West Bengal

Cluster Analysis for Terai Agro-ecological zone of West Bengal

As derived from the cluster analysis by using single linkage the whole of the respondents of Terai Zone here have been accommodated in 4 homogenous clusters.

Cluster 1 has accommodated 118 respondents, Cluster 2 has accommodated 9 respondents, Cluster 3 has accommodated 19 respondents and Cluster 4 has accommodated 8 respondents

The cluster has been formed because of the homogeneity of the respondents and distances between four clusters have been seen by measurable Euclidean distance. The respondents from alluvial zone can be segregated into five homogeneous groups. For CA socialization and entrepreneurship, each one of these groups can act as **Peer Entrepreneur on CA.**

Cluster 1 has accommodated 118 respondents, so they can form the **peer-1** with 118 CA farmers or, entrepreneur; **Cluster 2 has accommodated 9 respondents,** so they can form the **peer-2** with 09 CA farmers, or, entrepreneur, **Cluster 3 has accommodated 19 respondents,** so they can form the **peer-3** with 19 CA farmers, or, entrepreneur; **Cluster 4 has accommodated 8 respondents;** so they can form the **peer-4 with 8 CA farmers as well as entrepreneurs**

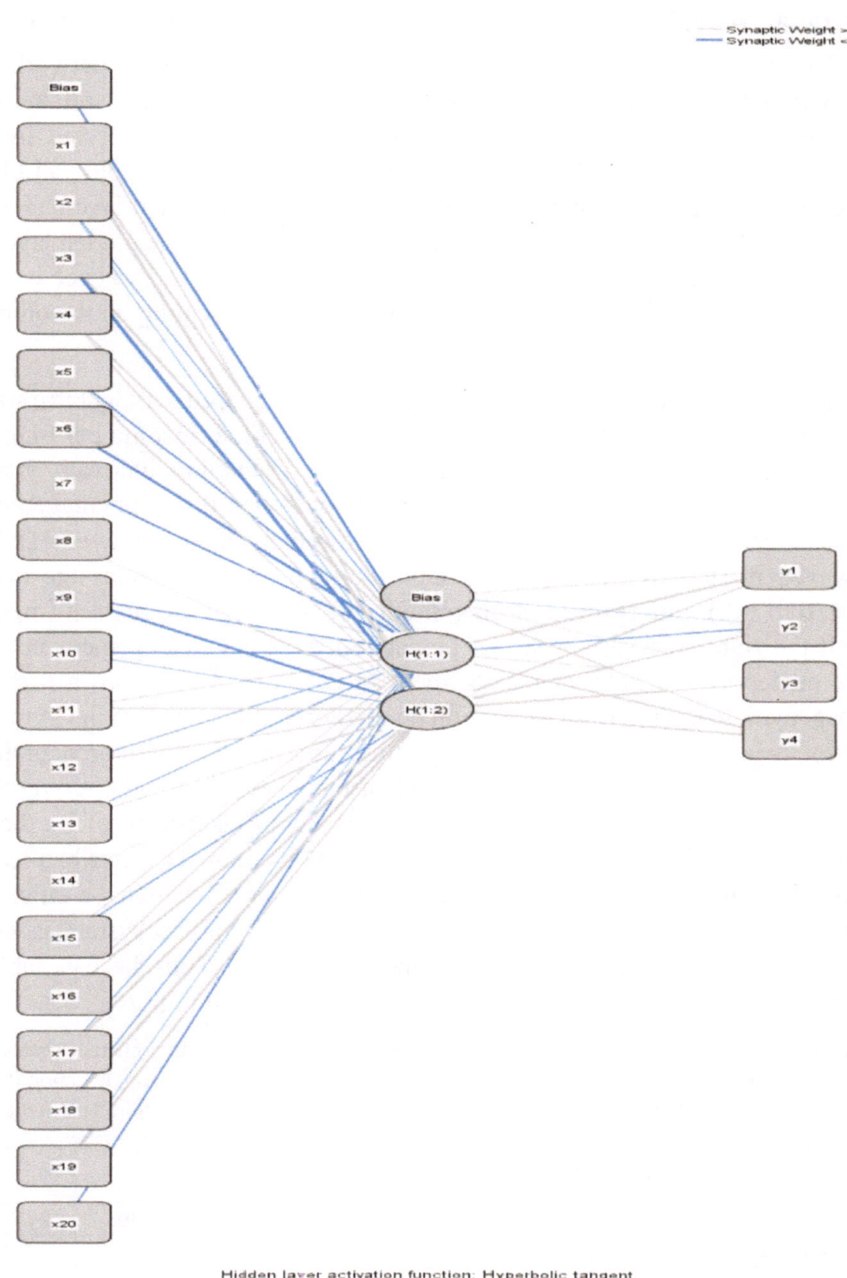

Fig. Artificial Neural Network of interaction on independent variables with dependent variables in Terai Zone of West Bengal

Artificial Neural Network Analysis for Terai Zone of West Bengal

ANN is the mimicry of human brain in analyzing the complex network of interaction between two sets of variables viz. input variables and output variables. Out of these polyhedral interactions, some of the selected input variables have been found to generate a swashbuckling effect (bold and blue lines) on the output variables after passing through two hidden layers (H1:1 and H1:2). When these inputs variables are passing through the hidden layers, error has been rationalized to the extent possible and what we get is the substantive effect of input variables in characterizing or scaling up the output variables.

From the set of input variables the following variables age, education, farm size, cropping intensity, number of fragments, annual income, income per unit of land, annual expenditure, stubble height, scientific orientation, innovativeness, extension agency contact, information seeking behavior, residue management score and mass media utilization have passed through the hidden layer (H1:1). However, in steering the impact of input variable from left side, **age (x1), education (x2) and residue management (x17)** has got the highest impact on all the right side variables, viz. perception on water management in CA (y1), perception on nutrient management in CA (y2), perception in carbon management in CA (y3) and perception on ecosystem management in CA (y4). The young generation farmers in terai region are more in to farm entrepreneurship and resource management. With an already established system of CA over the last decade, residue management is efficiently being taken care of by these young farmers. Education and training on CA and its related aspects had led a tremendous impact on the young minds which empowered them to be efficient in managing water, nutrient, soil organic carbon and overall ecosystem as well which has been reflected through the delineation of ANN. Thus these variables has an impact on perception change of the farmers of terai region.

Table 27: Descriptive statistics of the socio-economic and communication characteristics of the respondents in terms of Range (Minimum and Maximum), Mean, Standard Deviation (SD) and Coefficient of Variation (CV) of both alluvial and terai zone (pooled data) in West Bengal

Variables	Range		Mean	SD	CV (%)
	Minimum	Maximum			
Age of the respondent (x_1)	25.00	82.00	50.10	12.305	24.56
Education (x_2)	1.00	6.00	3.53	1.431	40.56
Family size (x_3)	1.00	22.00	5.55	2.776	50.05
Farm size (x_4)	1.50	40.00	12.12	8.814	72.70
Cropping intensity (x_5)	72.00	450.00	244.94	65.609	26.79
Number of land fragments (x_6)	1.00	22.00	3.80	3.935	103.45

Variables	Range		Mean	SD	CV (%)
	Minimum	Maximum			
Annual income (x_7)	25000.00	1540000.00	199308.08	149065.07	74.79
Income per capita (x_8)	5700.00	366250.00	37073.32	29972.014	80.84
Income per unit of land (x_9)	8303.13	213333.33	33666.87	25488.309	75.71
Annual expenditure (x_{10})	12000.00	624500.00	113426.81	79025.896	69.67
Stubble height (x_{11})	1.00	28.00	6.88	3.931	57.10
Volume of residue (x_{12})	325.00	26050.00	8285.39	6624.474	79.95
Scientific orientation (x_{13})	11.00	30.00	19.61	3.535	18.02
Innovativeness (x_{14})	7.00	23.00	15.07	3.184	21.14
Extension agency contact (x_{15})	2.00	16.00	8.58	3.391	39.52
Information seeking behavior (x_{16})	5.00	18.00	12.45	2.681	21.53
Residue management score (x_{17})	4.00	19.00	11.97	2.672	22.32
Perception on natural resource degradation (x_{18})	33.75	87.50	65.30	10.175	15.58
Number of livestock owned (cattle) (x_{19})	1.00	15.00	3.38	2.382	70.48
Mass media utilization (x_{20})	1.00	9.00	5.54	1.829	33.04
Perception on water management in CA (y_1)	4.00	16.00	12.63	2.512	19.89
Perception on nutrient management in CA (y_2)	8.00	16.00	13.59	1.903	14.01
Perception on carbon management in CA (y_3)	4.00	16.00	11.17	2.414	21.61
Perception on ecosystem management in CA (y_4)	6.84	15.66	12.35	1.985	16.08

Table 27 represents the distribution of variables in terms of range i.e. maximum and minimum values, mean, standard deviation and coefficient of variation for alluvial and terai zone (pooled data) of West Bengal.

It has been found that for the independent variable **age (x_1)**, the range lies between 82 and 25 years, where the oldest respondent is of 82 years and the youngest respondent being 25 years old. The mean value of the age of the respondents in this study was found to be 50.10 years with a standard deviation of 12.305 in the dataset. Coefficient of variation for this variable was found to be 24.56 per cent, which indicates a high level of consistency in the distribution.

It has been found that for the independent variable **education (x_2)**, the range lies between the scale value 1 and 6, which means the minimum qualification amongst the respondents is illiteracy and the highest formal education received among the respondents is either a graduate or post graduate. The mean scale value lies at 3.53, which means majority of the respondents received formal education up to middle school. The standard deviation in the data set is measured at 1.431. Coefficient of variation for this variable was found to be 40.56 per cent, which indicates the level of consistency in the distribution is medium.

It has been found that for the independent variable **family size (x_3)**, the range stretches within a minimum of 1 and a maximum of 22 family members. The average size of family consists of approximately 6 (5.55) members. The measured standard deviation in the dataset was found to be 2.776 and the calculated coefficient of variation is 50.05 per cent. This indicates there is a moderate to high level of consistency in distribution of data.

It has been found that for the independent variable **farm size (x_4)**, the range stretches between minimum of 1.5 bighas and maximum of 40 bighas of land. The mean value of farm size of the respondents in this study was found to be 12.12 bighas with a standard deviation of 8.814 in the dataset. Coefficient of variation for this variable was found to be 72.70 per cent, which indicates a moderate level of consistency in the distribution.

It has been found that for the independent variable **cropping intensity (x_5)**, the minimum was observed at 72.00 per cent and maximum at 450.00 per cent. The mean value of cropping intensity in this study stands at 244.94 per cent with a standard deviation of 65.609 in the dataset. Coefficient of variation observed for this variable was 26.79 per cent, which indicates a high level of consistency in the distribution.

It has been found that for the independent variable **number of land fragments (x_6)**, the range lies between 1 (minimum) and 22 (maximum). The mean value of number of land fragments observed in this study was 3.80 with a standard deviation of 3.935. Coefficient of variation for this variable was 103.45 per cent, which indicates inconsistency in the distribution.

It has been found that for the independent variable **annual income (x_7)**, the minimum is 25000 INR and the maximum is 1540000 INR. The mean value is 199308.08 INR with a standard deviation of 149065.07 in the dataset. Coefficient of variation for this variable was found to be 74.79 per cent, which indicates a low level of consistency in the distribution.

It has been found that for the independent variable **income per capita (x_8)**, the minimum is 5700 INR and the maximum is 366250 INR. The mean value is 37073.32 INR and the standard deviation is 29972.01. Coefficient of variation for this variable was found to be 80.84 per cent, which indicates a low of consistency in the distribution.

For the independent variable **income per unit of land (x_9)**, it has been found the minimum is 8303.13 INR and the maximum is 213333.33 INR. The mean value is 33666.87 with a standard deviation of 25488.31 in the total distribution. Coefficient of variation for this variable was found to be 75.71 per cent, which indicates a medium level of consistency in the distribution.

It has been found that for the independent variable **annual expenditure (x_{10})**, the minimum expenditure is 12000 INR and the maximum is 624500 INR. The mean value is 113426.81 INR with a standard deviation of 79025.896 in the total distribution. Coefficient of variation estimated at 69.67 per cent, which indicates a medium level of consistency in the distribution.

It has been found that for the independent variable **stubble height retention (x_{11})**, the minimum height maintained is 1 inch and the maximum is 28 inches. The mean height is 6.83inches with a standard deviation of 3.931 in total distribution. Coefficient of variation for this variable was found to be 57.10 per cent, which indicates a moderate level of consistency in the distribution.

It has been found that for the independent variable **volume of residue generated on field (x_{12})**, the minimum amount of residue generated in a farmer's field was 325 kgs and the maximum amount was 26050 kgs. The mean amount was 8285.39 with a standard deviation of 6624.474 in the total distribution. Coefficient of variation for this variable was 79.95 per cent, which indicates a medium level of consistency in the distribution.

It has been found that for the independent variable **scientific orientation (x_{13})**, the minimum score is 11 and the maximum score obtained is 30. The mean score lies at 19.61 with a standard deviation of 3.54 in the total distribution. Coefficient of variation for this variable was found to be 18.02 per cent, which indicates a very high level of consistency in the distribution.

It has been found that for the independent variable **innovativeness (x_{14})**, the minimum score is 7 and the maximum score obtained is 23. The mean score lies at 15.07 with a standard deviation of 3.184 in the total distribution. Coefficient of variation for this variable was found to be 21.14 per cent, which indicates a high level of consistency in the distribution.

It has been found that for the independent variable **extension agency contact (x_{15})**, the minimum score is 2 and the maximum score obtained is 16. The mean score lies at 8.58 with a standard deviation of 3.391 in the total distribution. Coefficient of variation for this variable was found to be 39.52 per cent, which indicates a high level of consistency in the distribution.

It has been found that for the independent variable **information seeking behaviour (x_{16})**, the minimum score is 5 and the maximum score obtained is 18. The mean score lies at 12.454 with a standard deviation of 2.681 in the total distribution. Coefficient of variation for this variable was found to be 21.53 per cent, which indicates a very high level of consistency in the distribution.

It has been found that for the independent variable **residue management (x_{17})**, the minimum score is 4 and the maximum score obtained is 19. The mean score lies at 11.97 with a standard deviation of 2.672 in the total distribution. Coefficient of variation for this variable was found to be 22.32 per cent, which indicates a very high level of consistency in the distribution.

It has been found that for the independent variable **perception on natural resource degradation (x18)**, the minimum score is 33.75 and the maximum score obtained is 87.50. The mean score lies at 65.30 with a standard deviation of 10.175 in the total distribution. Coefficient of variation for this variable was found to be 15.583 per cent, which indicates a high level of consistency in the distribution.

It has been found that for the independent variable **number of livestock owned (x_{19})**, the minimum count is 1 and the maximum is 15. The mean count is approximately 3 (3.380) with a standard deviation of 2.382 in the total distribution. Coefficient of variation for this variable was found to be 70.48 per cent, which indicates a medium level of consistency in the distribution.

It has been found that for the independent variable **mass media utilization (x_{20})**, the minimum score is 1 and the maximum score obtained is 9. The mean score lies at 5.54 with a standard deviation of 1.83 in the total distribution. Coefficient of variation for this variable was found to be 33.03 per cent, which indicates a high level of consistency in the distribution.

It has been found that for the dependent variable **perception on water management in CA (y_1)**, the minimum score is 4 and the maximum score obtained is 16. The mean score lies at 12.63 with a standard deviation of 2.512 in the total distribution. Coefficient of variation for this variable was found to be 19.89 per cent, which indicates a very high level of consistency in the distribution.

It has been found that for the dependent variable **perception on nutrient management in CA (y_2)**, the minimum score is 8 and the maximum score obtained is 16. The mean score lies at 13.59 with a standard deviation of 1.903 in the total distribution. Coefficient of variation for this variable was found to be 14.01 per cent, which indicates a very high level of consistency in the distribution.

It has been found that for the dependent variable **perception on carbon management in CA (y_3)**, the minimum score is 4 and the maximum score obtained is 16. The mean score lies at 11.17 with a standard deviation of 2.414 in the total distribution. Coefficient of variation for this variable was found to be 21.61 per cent, which indicates a high level of consistency in the distribution.

It has been found that for the dependent variable **perception on ecosystem management in CA (y_4)**, the minimum score is 6.84 and the maximum score obtained is 15.66. The mean score lies at 12.35 with a standard deviation of 1.985 in the total distribution. Coefficient of variation for this variable was found to be 16.08 per cent, which indicates a very high level of consistency in the distribution

Table 28: Correlation Coefficient of perception on water management (y_1) vs. 20 independent variables (x_1-x_{20}) for pooled data (AZ and TZ)

Independent Variables	'r' Value	Remarks
Age of the respondent (x_1)	0.195	**
Education (x_2)	0.270	**
Family size (x_3)	-0.095	
Farm size (x_4)	0.089	
Cropping intensity (x_5)	0.260	**
Number of land fragments (x_6)	-0.054	
Annual income (x_7)	0.109	
Income per capita (x_8)	0.046	
Income per unit of land (x_9)	-0.126	
Annual expenditure (x_{10})	0.099	
Stubble height (x_{11})	0.317	**
Volume of residue (x_{12})	0.059	
Scientific orientation (x_{13})	0.249	**
Innovativeness (x_{14})	0.290	**
Extension agency contact (x_{15})	-0.068	
Information seeking behavior (x_{16})	0.062	
Residue management score (x_{17})	-0.022	
Perception on natural resource degradation (x_{18})	-0.004	
Number of livestock (x_{19})	-0.143	*
Mass media utilization (x_{20})	0.168	*

*at 5% level of significance

**at 1% level of significance

Table 28 presents the correlation coefficient of correlation between perception on water management (y_1) and 20 independent variables (x_1-x_{20}). It is discernable from the table that from the variables, **age, education, cropping**

intensity, stubble height, scientific orientation, innovativeness and **mass media utilization** have recorded positive and significant correlation with the dependent variable under discussion. The variable **number of livestock** recorded a negative but significant correlation with the same dependent variable under study.

This findings is supported by the revelations made by **Ntshangase et al. (2018)** confirms that education plays a vital role in perception development that ultimately culminate into adoption of no-till practices. Stubble height helps in moisture retention. With more exposure to the outer world, better perception development on water management is justified.

Table 29: Path analysis of perception on water management in CA (y_1) vs. 20 exogenous variables ($x_1 - x_{20}$) variables in Alluvial and terai zone (pooled) of West Bengal (n=229)

Exogenous Variables	Total Effect	Direct Effect	Indirect Effect	Highest Indirect Effect
Age of the respondent (x_1)	0.195	0.265	-0.070	-0.028(x_5)
Education (x_2)	0.270	0.213	0.057	0.043(x_{10})
Family size (x_3)	-0.095	-0.164	0.069	0.083(x_{10})
Farm size (x_4)	0.089	0.026	0.063	-0.045(x_3)
Cropping intensity (x_5)	0.260	0.200	0.060	-0.037(x_1)
Number of land fragments (x_6)	-0.054	-0.141	0.087	0.076(x_{10})
Annual income (x_7)	0.109	0.046	0.063	0.091(x_{10})
Income per capita (x_8)	0.046	-0.082	0.128	0.098(x_{10})
Income per unit of land (x_9)	-0.126	-0.175	0.049	0.018(x_6)
Annual expenditure (x_{10})	0.099	0.194	-0.095	-0.070(x_3)
Stubble height (x_{11})	0.317	0.210	0.107	0.033(x_{14})
Volume of residue (x_{12})	0.059	-0.030	0.089	0.036(x_{13})
Scientific orientation (x_{13})	0.249	0.165	0.084	0.040(x_{11})
Innovativeness (x_{14})	0.290	0.146	0.144	0.047(x_{11})
Extension agency contact (x_{15})	-0.068	-0.013	-0.055	0.068(x_6)
Information seeking behavior (x_{16})	0.062	0.015	0.047	-0.056(x_{10})
Residue management score (x_{17})	-0.022	-0.044	0.022	0.021(x_1)
Perception on natural resource degradation (x_{18})	-0.004	-0.073	0.069	0.033(x_1)
Number of livestock (x_{19})	-0.143	-0.132	-0.011	0.039(x_6)
Mass media utilization (x_{20})	0.168	0.062	0.106	0.044(x_{13})

Residual effect: 0.590

Table 29 presents the path analysis of dependent variable perception on water management (y_1) wherein the total effect (coefficient of correlation) has been decomposed into direct, indirect and residual effect.

The variable age has exerted the highest direct effect on perception on water management. CA is going to be the lucrative proposition for the young age farmers. So school dropouts, graduate entrepreneurs, natural ecologists, students (school girls and boys) can be selected for the robust socialization of CA, not to produce grain only but to sustain ecology as well.

The variable innovativeness has exerted highest indirect effect on on perception on water management. Innovation and experiments are the prime mover for the success of CA. in this approach, we need to transform all non-cognitive adoptions into conviction of ecological skill and perception.

Whatever may be the ecological benefits, it will be futile unless income and expenditure incur with CA are prudently taken care. At the end of the day, every farmers, and they have the right to think so, a happy return from their investment to defend their family expenditure and yeomen services. That is why the variable annual expenditure has routed highest indirect individual effect on as many as 6 exogenous variables.

The residual effect being 0.590 means that 59% of the variance couldn't been explained with these combination of 20 exogenous variables.

Table 30: Stepwise regression analysis of perception on water management in CA (y_1) versus 20 independent ($x_1 - x_{20}$) variables in alluvial and terai zone (pooled) of West Bengal (n=229)

Variables	Reg. coeff. B	S.E. B	Beta	t value	R^2	Std. error of the estimate
Stubble height (x_{11})	0.141	0.036	0.219	3.925	0.367	2.039
Education (x_2)	0.415	0.096	0.236	4.341		
Age (x_1)	0.052	0.011	0.255	4.684		
Cropping intensity (x_5)	0.008	0.002	0.211	3.825		
Scientific orientation (x_{13})	0.127	0.040	0.179	3.204		
Innovativeness (x_{14})	0.131	0.044	0.166	2.968		
No. of livestock owned (cattle) (x_{19})	-0.150	0.057	-0.142	-2.639		
Income per unit of land (x_9)	0.000	0.000	-0.131	-2.428		

Table 30 shows that variables, **stubble height, education, age, cropping intensity, scientific orientation, innovativeness, number of livestock owned and income per unit of land** have been retained at the last step. The R^2 value being 0.367, reveals that these three variables together explains 36.7 per cent of the variance embedded in **perception on water management in CA (y_1)** in terai region of West Bengal.

To learn advanced and sustainable farming technology, education is a key element. Proper knowledge and adequate support from the extension agents helps to in developing scientific outlook and innovativeness in the farmers. This changed outlook would help in perceptional change of the farmers towards water management in their respective farms. Also cropping intensity is a determining factor in perceptional development in farmers as those who farm intensively will make more efficient use of resources in comparison to those who till once or twice a year.

Table 31: Correlation Coefficient of perception on nutrient management (y_2) vs. 20 independent variables (x_1-x_{20}) of pooled data (AZ and TZ)

Independent Variables	'r' Value	Remarks
Age of the respondent (x_1)	0.304	**
Education (x_2)	0.230	**
Family size (x_3)	-0.097	
Farm size (x_4)	0.332	**
Cropping intensity (x_5)	-0.024	
Number of land fragments (x_6)	0.091	
Annual income (x_7)	0.178	**
Income per capita (x_8)	0.142	*
Income per unit of land (x_9)	-0.015	
Annual expenditure (x_{10})	0.114	
Stubble height (x_{11})	0.090	
Volume of residue (x_{12})	0.377	**
Scientific orientation (x_{13})	0.051	
Innovativeness (x_{14})	0.110	
Extension agency contact (x_{15})	-0.101	
Information seeking behavior (x_{16})	0.061	
Residue management score (x_{17})	0.260	**
Perception on natural resource degradation (x_{18})	0.265	**
Number of livestock (x_{19})	-0.047	
Mass media utilization (x_{20})	0.058	

*at 5% level of significance
**at 1% level of significance

Table 31 presents the correlation coefficient of correlation between perception on nutrient management (y_2) and 20 independent variables (x_1-x_{20}). It is discernable from the table that from the variables, age, education, farm size, annual income, income per capita, volume of residue, residue management score and perception on natural resource degradation have recorded positive and significant correlation with the dependent variable, perception on nutrient management under discussion.

On farm nutrient management is an important sustainable practice. Using farm residues as nutrient sources requires experiences that comes with age of practice. More the residue generated on farm, better the management of the residues, more the knowledge on natural resource degradation, it is quite natural to develop a better perception on nutrient management among the farmers. Bigger farm size helps in more residue generation, which when managed properly helps in management of nutrient application as well. Experience, education and income also plays vital role in perception building of an individual.

Table 32: Path analysis of perception on nutrient management in CA (y_2) vs. 20 exogenous variables ($x_1 - x_{20}$) variables in Alluvial and terai zone (pooled) of West Bengal (n=229)

Exogenous Variables	**Total Effect (TE)**	**Direct Effect (DE)**	**Indirect Effect (IE)**	**Highest Indirect Individual Effect (HIIE)**
Age of the respondent (x_1)	0.304	0.219	0.085	**0.034(x_{12})**
Education (x_2)	0.230	0.109	0.121	0.049(x_4)
Family size (x_3)	-0.097	-0.228	0.131	0.065(x_4)
Farm size (x_4)	0.332	0.240	0.092	**0.103(x_{12})**
Cropping intensity (x_5)	-0.024	-0.078	0.054	-0.031(x_1)
Number of fragments (x_6)	0.091	0.041	0.050	0.086(x_{15})
Annual income (x_7)	0.178	0.045	0.133	**0.051(x_{12})**
Income per capita (x_8)	0.142	-0.042	**0.184**	0.042(x_3)
Income per unit of land (x_9)	-0.015	0.036	-0.051	**-0.035(x_{12})**
Annual expenditure (x_{10})	0.114	0.079	0.035	-0.098(x_3)
Stubble height (x_{11})	0.090	0.074	0.016	0.032(x_4)
Volume of residue (x_{12})	0.377	**0.290**	0.087	0.085(x_4)
Scientific orientation (x_{13})	0.051	-0.052	0.103	**0.063(x_{12})**
Innovativeness (x_{14})	0.110	0.048	0.062	0.027(x_4)
Extension agency contact (x_{15})	-0.101	-0.177	0.076	**0.071(x_{12})**
Information seeking behavior (x_{16})	0.061	0.029	0.032	-0.073(x_{15})
Residue management score (x_{17})	0.260	0.164	0.096	**0.038(x_{12})**
Perception on natural resource degradation (x_{18})	0.265	0.129	0.136	**0.064(x_{12})**
Number of livestock (x_{19})	-0.047	-0.112	0.065	-0.068(x_{15})
Mass media utilization (x_{20})	0.058	0.029	0.029	-0.021(x_{15})

Residual effect: 0.568

Table 32 presents the path analysis of dependent variable perception on nutrient management (y_2) wherein the total effect (coefficient of correlation) has been decomposed into direct, indirect and residual effect.

For both the zones, the volume of residue management has generated the conspicuous effect on nutrient mgt. through harvesting of crops, the soil ecology suffers huge amount of nutrient removal that need to be replenished and compensated by mixing crop residues in the soil. These crop residues are the source of NPK, OC, minerals and even microorganisms. That is how a policy framework should be there not a kg of CR cannot be removed from the field are destroyed through burning either.

Every bit of intervention made in the farm should have an income implication unless the CA approaches an execution result into happy return, it would be challenging and difficult for retention and further upscaling of CA.

The residual effect being 0.568 means that 56.8% of the variance couldn't been explained with these combination of 20 exogenous variables.

Table 33: Stepwise regression analysis of perception on nutrient management in CA (y_2) versus 20 independent ($x_1 - x_{20}$) variables in alluvial and terai zone (pooled) of West Bengal (n=229)

Variables	Reg. coeff. B	S.E. B	Beta	t value	R^2	Std. error of the estimate
Volume of residue (x_{12})	0.000	0.000	0.282	4.925	0.411	1.493
Age (x_1)	0.036	0.008	0.230	4.315		
Education (x_2)	0.173	0.074	0.130	2.347		
Residue management score (x_{17})	0.112	0.038	0.157	2.940		
Farm size (x_4)	0.056	0.013	0.261	4.230		
Extension agency contact (x_{15})	-0.100	0.034	-0.177	-2.931		
Family size (x_3)	-0.120	0.038	-0.175	-3.188		
Perception on natural resource degradation (x_{18})	0.026	0.011	0.138	2.430		
No. of livestock owned (cattle) (x_{19})	-0.108	0.046	-0.134	-2.343		

Table 33 shows that variables, **volume of residue, age, education, residue management score, farm size, extension agency contact, family size, perception on natural resource degradation,** and **number of livestock** have been retained at the last step. The R^2 value being 0.411, reveals that these two variables together explains 41.1 per cent of the variance embedded in **perception on nutrient management in CA (y_2)** in terai region of West Bengal.

Perception on natural resource degradation has got a direct impact on managing natural resources like residues for managing soil nutrition. With bigger farms and more residues, this gives ample scope to farmers to manage nutrients at farm recycling of the residues.

Table 34: Correlation Coefficient of perception on carbon management (y_3) vs. 20 independent variables (x_1-x_{20}) of pooled data (AZ and TZ)

Independent Variables	'r' Value	Remarks
Age of the respondent (x_1)	0.058	
Education (x_2)	0.301	**
Family size (x_3)	-0.103	
Farm size (x_4)	0.051	
Cropping intensity (x_5)	0.201	**
Number of land fragments (x_5)	0.209	**
Annual income (x_7)	0.193	**
Income per capita (x_8)	0.076	
Income per unit of land (x_9)	-0.014	
Annual expenditure (x_{10})	0.081	
Stubble height (x_{11})	0.195	**
Volume of residue (x_{12})	0.126	
Scientific orientation (x_{13})	0.317	**
Innovativeness (x_{14})	0.087	
Extension agency contact (x_{15})	-0.063	
Information seeking behavior (x_{16})	0.051	
Residue management score (x_{17})	0.117	
Perception on natural resource degradation (x_{18})	0.114	
Number of livestock (x_{19})	-0.177	**
Mass media utilization (x_{20})	0.419	**

*at 5% level of significance
**at 1% level of significance

Table 34 presents the correlation coefficient of correlation between perception on carbon management (y_3) and 20 independent variables (x_1-x_{20}). It is discernable from the table that from the variables, **education, cropping intensity, number of fragments, annual income, stubble height, scientific orientation,** and **mass media utilization** have recorded positive and significant correlation and the variable number of livestock has recorded negative but significant correlation with the dependent variable, perception on carbon management (y_3) under discussion.

Education helps to build up proper understanding on ecological surrounding and function. The new generation, young farmers have already been exposed, in their school pedagogy or training experiences on organic farming vis-à-vis CA, are more prone to protecting ecological aspects of farming. This offers a silver line for the success of CA.

Higher level of cropping intensity and inclusion of eco-friendly crops in the intercropping practices, the prospect of ecological management including soil, water and biodiversity are gaining stronger grounds. High intensity farming, after attaining highest possible productivity per unit area, shows a natural propensity to go after CA to restore whatever has already been depleted to support high yield behavior of farming.

Table 35: Path analysis of perception on carbon management in CA (y_3) vs. 20 exogenous variables ($x_1 - x_{20}$) variables in Alluvial and terai zone (pooled) of West Bengal (n=229)

Exogenous Variables	Total Effect	Direct Effect	Indirect Effect	Highest Indirect Effect
Age of the respondent (x_1)	0.058	0.109	-0.051	-0.020(x_{13})
Education (x_2)	0.301	0.238	0.063	**0.030(x_6)**
Family size (x_3)	-0.103	-0.130	0.027	**0.033(x_6)**
Farm size (x_4)	0.051	-0.030	0.081	0.049(x_2)
Cropping intensity (x_5)	0.201	0.111	0.090	0.027(x_2)
Number of land fragments (x_6)	0.209	0.178	0.031	0.047(x_{19})
Annual income (x_7)	0.193	0.074	0.119	0.040(x_2)
Income per capita (x_8)	0.076	-0.027	0.103	0.041(x_2)
Income per unit of land (x_9)	-0.014	-0.027	0.013	0.032(x_{20})
Annual expenditure (x_{10})	0.081	0.025	0.056	**0.070(x_6)**
Stubble height (x_{11})	0.195	0.117	0.078	0.058(x_{20})
Volume of residue (x_{12})	0.126	0.054	0.072	0.040(x_{13})
Scientific orientation (x_{13})	0.317	0.186	0.131	0.085(x_{20})
Innovativeness (x_{14})	0.087	-0.075	**0.162**	0.061(x_{20})
Extension agency contact (x_{15})	-0.063	0.037	-0.100	**-0.086(x_6)**
Information seeking behavior (x_{16})	0.051	0.037	0.014	**-0.050(x_6)**
Residue management score (x_{17})	0.117	0.020	0.097	0.033(x_{20})
Perception on natural resource degradation (x_{18})	0.114	0.091	0.023	-0.036(x_{19})
Number of livestock (x_{19})	-0.177	-0.170	-0.007	**-0.049(x_6)**
Mass media utilization (x_{20})	0.419	**0.319**	0.100	0.049(x_{13})

Residual effect: 0.578

Table 35 presents the path analysis of dependent variable perception on carbon management (y_3) wherein the total effect (coefficient of correlation) has been decomposed into direct, indirect and residual effect.

The variable **scientific orientation** has exerted **highest direct effect** on perception on carbon management. To know how to manage carbon, one must know the why at first. This why is the science behind the importance of managing organic carbon on farm. Unless one knows the contribution of managing carbon to his farm, it is difficult to convince farmers to do so since contribution of soil organic carbon is not readily visible to the farmers.

The variable **innovativeness** has exerted **highest indirect effect** on perception on carbon management (y_3). Innovativeness is necessary to be keen on applying new practices on field. An innovative farmer, out of his curiosity when communicate with different sources to gather knowledge on certain practices, his perception on that topic gradually develops.

It has been found that variable **number of fragments** has routed **highest indirect individual effect** on as many as **6 exogenous variables.**

The residual effect being 0.578 means that 57.8% of the variance couldn't been explained with these combination of 20 exogenous variables.

Table 36: Stepwise regression analysis of perception on carbon management in CA (y_3) versus 20 independent ($x_1 - x_{20}$) variables in alluvial and terai zone (pooled) of West Bengal (n=229)

Variables	Reg. coeff. B	S.E. B	Beta	t value	R^2	Std. error of the estimate
Mass media utilization (x_{20})	0.442	0.073	0.334	6.085	0.388	1.927
Education (x_2)	0.423	0.092	0.250	4.605		
Scientific orientation (x_{13})	0.151	0.038	0.220	3.931		
No. of livestock owned (cattle) (x_{19})	-0.159	0.057	-0.157	-2.804		
No. of fragments (x_6)	0.085	0.035	0.138	2.443		
Perception on natural resource degradation (x_{18})	0.029	0.013	0.123	2.217		
Cropping intensity (x_5)	0.005	0.002	0.125	2.314		
Age (x_1)	0.022	0.011	0.112	2.065		

Table 23 shows that variables, **mass media utilization, education, scientific orientation, number of livestock owned, number of fragments, perception on natural resource degradation, cropping intensity and age** have been retained at the last step. The R^2 value being 0.388, reveals that these retained variables together explains 38.8 per cent of the variance embedded in **perception on carbon management in CA (y_3)** in terai region of West Bengal.

All the variables retained at the last have important contribution in perception building. Education, access to mass media and scientific orientation plays major role in perception development of a person. Cattle produce manures which are rich source of organic carbon when decomposed and transformed into manures.

Table 37: Correlation Coefficient of perception on ecosystem management in CA (y_4) vs. 20 independent variables (x_1-x_{20}) for pooled data (AZ & TZ)

Independent Variables	'r' Value	Remarks
Age of the respondent (x_1)	0.143	*
Education (x_2)	0.332	**
Family size (x_3)	-0.137	*
Farm size (x_4)	0.076	
Cropping intensity (x_5)	0.262	**
Number of land fragments (x_6)	0.072	
Annual income (x_7)	0.174	**
Income per capita (x_8)	0.066	
Income per unit of land (x_9)	-0.096	
Annual expenditure (x_{10})	0.090	
Stubble height (x_{11})	0.299	**
Volume of residue (x_{12})	0.102	
Scientific orientation (x_{13})	0.339	**
Innovativeness (x_{14})	0.272	**
Extension agency contact (x_{15})	-0.062	
Information seeking behavior (x_{16})	0.069	
Residue management score (x_{17})	0.072	
Perception on natural resource degradation (x_{18})	0.064	
Number of livestock (x_{19})	-0.182	**
Mass media utilization (x_{20})	0.336	**

*at 5% level of significance
**at 1% level of significance

Table 37 presents the coefficient of correlation between perception on ecosystem management in conservation agriculture (y_4) and 20 independent (x_1 - x_{20}) variables. It has been found that variables viz., **age, education, cropping intensity, annual income, stubble height, scientific orientation, innovativeness** and **mass media utilization** have recorded positive and significant correlation with the dependent variable under study. The variable **family size and number of livestock** showed a negative but significant correlation with the dependent variable under study.

Education and experience helps to build up proper understanding on ecological surrounding and function. Proper education aided with mass media helps to generate scientific outlook and instinct innovativeness in a person. This leads to better perception development in an individual. Ecosystem is an interaction among all the biotic and abiotic components in the nature. So, when an individual starts managing small elements of ecosystem, it starts managing ecosystem as a whole.

Table 38: Path analysis of perception on ecosystem management in CA (y_4) vs. 20 exogenous variables ($x_1 - x_{20}$) variables in Alluvial and terai zone (pooled) of West Bengal (n=229)

Exogenous Variables	Total Effect (TE)	Direct Effect (DE)	Indirect Effect (IE)	Highest Indirect individual Effect (HIIE)
Age of the respondent (x_1)	0.143	0.212	-0.069	-0.023(x_5)
Education (x_2)	0.332	**0.261**	0.071	**0.029(x_{10})**
Family size (x_3)	-0.137	-0.195	0.058	**0.057(x_{10})**
Farm size (x_4)	0.076	-0.011	0.087	0.054(x_2)
Cropping intensity (x_5)	0.262	0.169	0.093	0.030(x_2)
Number of land fragments (x_6)	0.072	0.012	0.060	**0.053(x_{10})**
Annual income (x_7)	0.174	0.072	0.102	**0.063(x_{10})**
Income per capita (x_8)	0.066	-0.064	0.130	**0.068(x_{10})**
Income per unit of land (x_9)	-0.096	-0.138	0.042	0.020(x_{20})
Annual expenditure (x_{10})	0.090	0.133	-0.043	-0.084(x_3)
Stubble height (x_{11})	0.299	0.185	0.114	0.040(x_{12})
Volume of residue (x_{12})	0.102	0.004	0.098	0.045(x_{12})
Scientific orientation (x_{13})	0.339	0.209	0.130	0.054(x_{20})
Innovativeness (x_{14})	0.272	0.104	**0.168**	0.042(x_{11})
Extension agency contact (x_{15})	-0.062	0.035	-0.097	-0.066(x_{19})
Information seeking behavior (x_{16})	0.069	0.011	0.058	-0.039(x_{19})
Residue management score (x_{17})	0.072	0.012	0.060	0.021(x_{20})
Perception on natural resource degradation (x_{18})	0.064	0.000	0.064	-0.036(x_{19})
Number of livestock (x_{19})	-0.182	-0.172	-0.010	**-0.026(x_{10})**
Mass media utilization (x_{20})	0.336	0.205	0.131	0.056(x_{13})

Residual effect: 0.523

Table 38 presents the path analysis of dependent variable perception on ecosystem management (y_4) wherein the total effect (coefficient of correlation) has been decomposed into direct, indirect and residual effect.

Exogenous variable **education** has exerted **highest direct effect** on perception on ecosystem management. Education promotes acquisition of skills and here in this case, skill in managing an ecosystem by a farmer. All the four basic functions, provisionary, regulatory and cultural services, supporting services needs proper education. That's how education turns out to be an important factor affecting the perception on ecosystem management.

Ecosystem management being complex and diverse needs seamless flow of innovations into its mentoring and regeneration. This justifies the **highest indirect effect** being exerted by the variable **innovativeness**.

The variable **annual expenditure** has routed **highest indirect individual effect** in as many as 6 exogenous variable.

The residual effect being 0.523 means that 52.3% of the variance couldn't been explained with these combination of 20 exogenous variables.

Table 39: Stepwise regression analysis of perception on ecosystem management in CA (y_4) versus 20 independent ($x_1 - x_{20}$) variables in alluvial and terai zone (pooled) of West Bengal (n=229)

Variables	Reg. coeff. B	S.E. B	Beta	t value	R2	Std. error of the estimate
Scientific orientation (x_{13})	0.128	0.030	0.227	4.276	0.463	1.491
Education (x_2)	0.405	0.070	0.291	5.773		
Mass media utilization (x_{20})	0.229	0.057	0.210	3.985		
Age (x_1)	0.034	0.008	0.212	4.194		
Cropping intensity (x_5)	0.006	0.002	0.183	3.579		
No. of livestock owned (cattle) (x_{19})	-0.153	0.042	-0.183	-3.680		
Stubble height (x_{11})	0.086	0.026	0.170	3.256		
Income per unit of land (x_9)	0.000	0.000	-0.124	-2.469		
Family size (x_3)	-0.081	0.036	-0.113	-2.256		
Innovativeness (x_{14})	0.066	0.033	0.106	2.023		

Table 39 shows that variables, **scientific orientation, education, mass media utilization, age, cropping intensity, number of livestock owned, stubble height, income per unit of land, family size** and **innovativeness** have been retained at the last step. The R^2 value being 0.463, reveals that these three variables together explains 46.3 per cent of the variance embedded in **perception on ecosystem management in CA (y_4)** in terai region of West Bengal.

Ecosystem management involves management of water, soil and biodiversity as well as eco-friendly entrepreneurship. All these demands a good scientific orientation of the practicing farmer. Education and exposure to mass media helps to develop the thought process generating into perception development. Retention of the variables confirms their substantive contribution in developing the perception on ecosystem management of the farmer.

Canonical Covariate Analysis for respondents of Alluvial and Terai zone (pooled) in West Bengal

CCA is another extension of multiple regression where rather than using a single outcome variable Y, two or more Y variables are predicted by two or more predictor X variables. The purpose is usually not to predict Ys from Xs but to explain the relationship between the X and Y variable sets. The analysis can be run in both directions, Xs predicting Ys and Ys predicting Xs, but the researcher is generally concerned about Xs predicting Ys where Y is the new hypothesized inter-relationship and the X set are the traditional predictors seen in prior research.

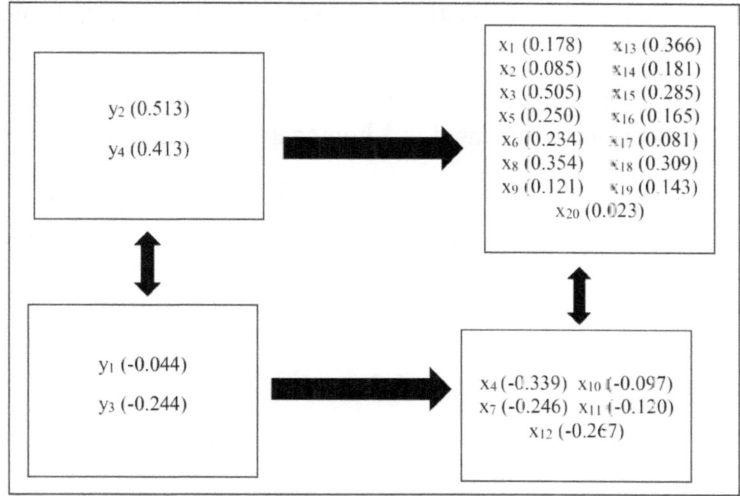

Fig. Canonical Covariate Analysis for the respondents in Alluvial and Terai Zone (pooled)

The CCA (pooled) model has been executed for the terai zone shows that there has been a sub conglomeration of two dependent variable perception on water management (y_1) and perception on carbon management (y_3). It may so happened that there might be a congenital bondage between perception on water management and perception on carbon management and when these are moving together, following variables from the right side variables set **farm size, annual income, annual expenditure, stubble height and volume of residue** are precisely influencing the depending variables. Similarly, for the left side dependent variables perception on nutrient management (y_2) and perception on ecosystem management (y_4) are moving together and when they are forming a homogenous agglomeration with the right side variables **age, education, family size, cropping intensity, number of fragments, income per capita, income per unit land, scientific orientation, innovativeness, extension agency contact, information seeking behavior, residue management score, perception on natural resource degradation, number of livestock owned and mass media utilization** are precisely influencing the dependent variables. Thus, this CCA model can help to take a well calibrated intervention or upscaling these four depending variables (left side variables). For CCA model, when we are considering the conjoint interactions between right and left side variables, then it has been found that pattern of interaction amongst and between the variables in the terai region is almost similar to that of CCA model generated on the pooled sample for which the alluvial zone is moving in a unique way.

Cluster Analysis for Alluvial and Terai Agro-ecological zone (pooled) of West Bengal

As derived from the cluster analysis by using single linkage the whole of the respondents here has been accommodated in 5 homogenous clusters.

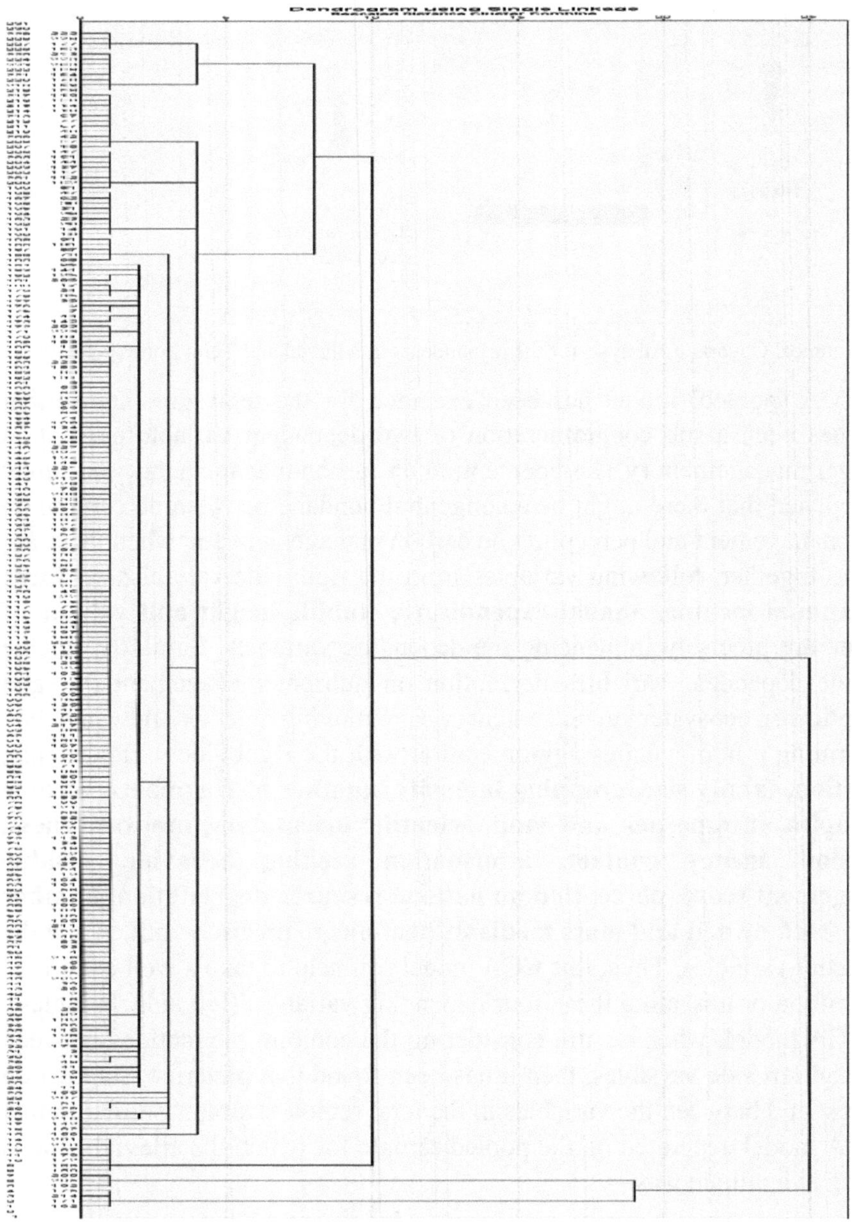

Fig. Cluster analysis for the respondents in Alluvial and Terai Zone (pooled)

Cluster 1 has accommodated 203 respondents, Cluster 2 has accommodated 7 respondents, Cluster 3 has accommodated 12 respondents, Cluster 4 has accommodated 4 respondents and Cluster 5 has accommodated 3 respondents.

The cluster has been formed because of the homogeneity of the respondents and distances between four clusters have been seen by measurable Euclidean distance .So, these 5 clusters can be considered as a strategic respondent group for taking any kind of strategic intervenes.

Cluster 1 has accommodated 203 respondents, so they can form the **peer-1** with 203 CA farmers or, entrepreneur; **Cluster 2 has accommodated 7 respondents,** so they can form the **peer-2** with 07 CA farmers, or, entrepreneur, **Cluster 4 has accommodated 04 respondents,** so they can form the **peer-4** with 19 CA farmers, or, entrepreneur; **Cluster 5 has accommodated 3 respondents;** so they can form the **peer-4 with 3 CA farmers as well as entrepreneurs.**

Artificial Neural Network Analysis of pooled data

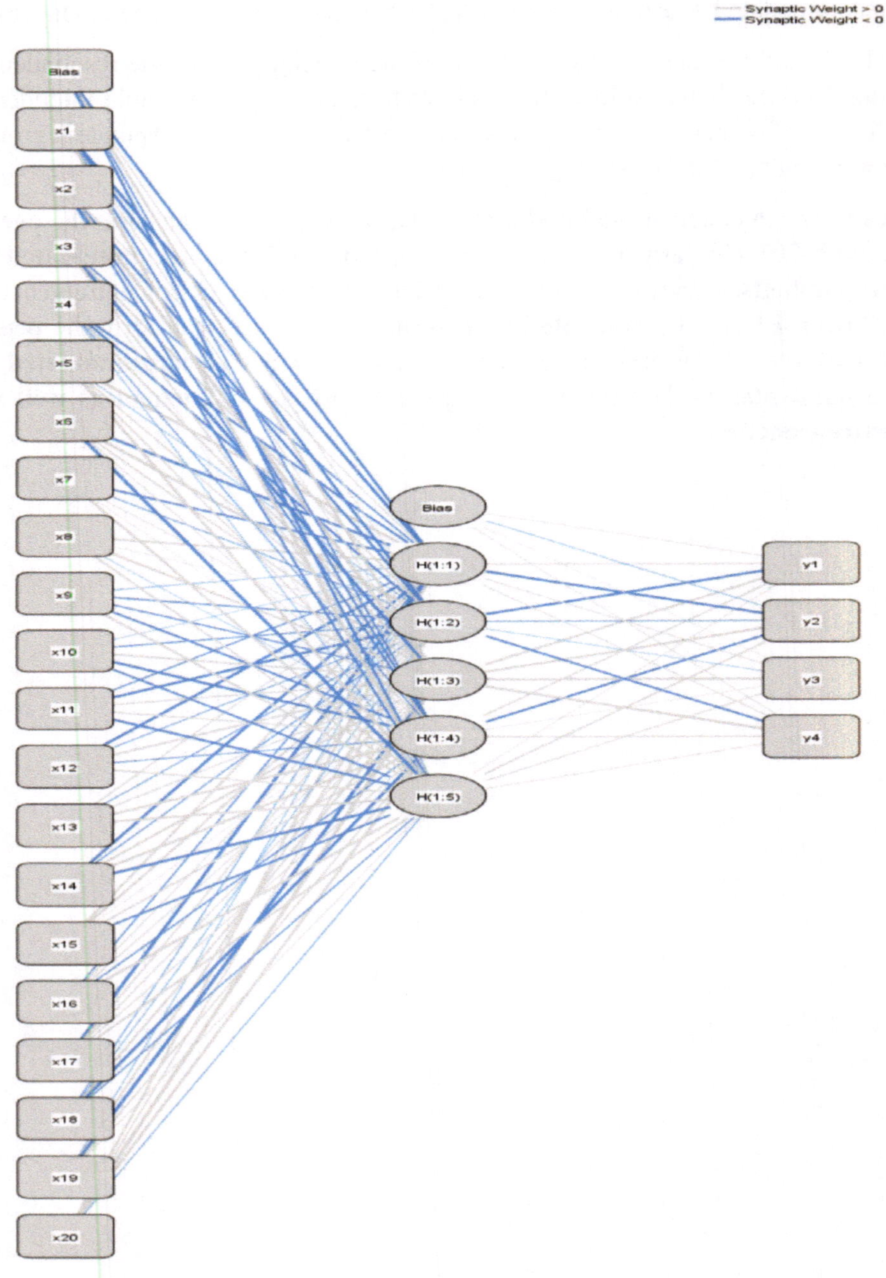

Fig. Artificial Neural Network Analysis for the respondents in Alluvial and Terai Zone (pooled)

ANN is the mimicry of human brain in analyzing the complex network of interaction between two sets of variables viz. input variables and output variables. Out of these polyhedral interactions, some of the selected input variables have been found to generate a swashbuckling effect (bold and blue lines) on the output variables after passing through five hidden layers (H1:1, H1:2, H1:3, H1:4 and H1:5). When these inputs variables are passing through the hidden layers, error has been rationalized to the extent possible and what we get is the substantive effect of input variables in characterizing or scaling up the output variables.

From the set of input variables the following variables x1, x3, x4, x6, x7, x9, x10, x11, x12, x14, x16 and x17 have passed through the hidden layer (H1:1). However, in steering the impact of input variable from left side, **age (x_1), volume of residue (x_{12}), innovativeness (x_{14}) and residue management (x_{17})** has got the highest impact on the right side variable, perception on nutrient management in CA (y_2). The overall nutrient management perception, combining both the zones, has been impacted by age, volume of residue, innovativeness…. So the following strategic implication coming out for both the agro climatic zones (AZ and TZ) are as follows

i. Young age farmers can be the harbinger for socialization of CA

ii. Volume of crop residues, whatever may it be, must be recycled in the operating farms to protect it from erosions, depletion and disruption

iii. Innovation is the prime mover for propagating CA. that's why in both the zones, factors like education has appeared directly or indirectly in several occasions. So, well designed training modules on CA, environmental stewardship, ecological entrepreneurship, community resource bank, soil-water-biodiversity stewardship need to be created for transforming the farmers from producers to mentor.

iv. Residue management is the crux of CA Cr are the gifts of agricultural production process and soil is the only proprietor to receive it. So, strong policy formulation need to be in place for creating crop residue bank in every village and befitting technology must be made available for the proper decomposition and mixing it with the soil without hampering collateral farm operations. Incentives and rewards for the CA farmers in both the zones can promote and motivate the farmers for effective crop residue management

Nothing can go more important than to restoration and reenergizing the ecological resilience else the civilization has to face an apocalyptic consequence.

Participatory Rural Appraisal (PRA)

The social ecology of this study area is undergoing moderate to fast transformation with elucidating impact on social life, institutional functions, naturalresource management, and livelihood. Participatory analysis of a social ecology has so far been complex, multidimensional, and collateral at the same time.

In the present study, SWOT analysis and ITK identification have been incorporated as participatory data generation approaches to visualize the community perception of Conservation Agriculture.

Steps to be followed during Participatory SWOT analysis

1. Formation of focus group comprising of local farmers, farm women, school teachers, and youth with different resource and experiential background.
2. Session 1 (Green session): All possible items are randomly extracted through facilitation.
3. Session 2 (Red session): Consensus based finalization of items.
4. Wisdom opinions are also collected.

Table 40: SWOT of Alluvial zone

Principles of CA	Strength	Weakness	Opportunity	Threats
Minimum soil disturbance	• Reduce cost of cultivation • Stops soil erosion	• Inherent belief on tillage practice • Clayey texture clogs machineries • Fragmented and disparity in land holding	• Improving soil health with enhanced productivity • Soil organic carbon status can be restored	• Resistance in acknowledging tillage benefits • Lack of curated farm machineries • Poor financial resources
Permanent soil cover	• Abundance of crop residues • Less competition in alternative uses	• Lack of skilled technicians to run machineries • Problem of termites in soil	• Increase in soil organic carbon pool • Proper fertilizer management	• Unwilling to compromise on 'extra income' • Burning saves money
Crop diversification	•Already diversified cropping system •Market is available	•Performance of every crop is still debatable	•Improved soil health and also income •Environment friendly and effective method to break pest and disease cycle	•Strict adherence to other principles might not apply to all crops

Table 41: SWOT of terai zone

Principles of CA	**Strength**	**Weakness**	**Opportunity**	**Threats**
Minimum soil disturbance	• Reduce cost of cultivation • Stops soil erosion • Profile of experience build up • Skilled tractor and seed drill driver	• Crop immersion in low land • Increasing land fragmentation	• Improving soil health with enhanced productivity • Soil organic carbon status can be restored • Practicing no-till for few years	• Heavy loan because of costly machinery purchase • Lack of timely availability of machinery • Natural calamities
Permanent soil cover	• Abundance of crop residues • Pulse crop production • Availability of maize Stover	• Livestock-CA competition for residues • Problem of termites in soil	• Increase in soil organic carbon pool • Proper fertilizer management • Lesser soil erosion	• Every household has at least one livestock • Labor problem
Crop diversification	• Already diversified cropping system • Introduction of pulse crop leads to better soil health • Support from FPO and SAU • Upcoming entrepreneurship • Market buoyancy for CA	• Performance of every crop is still debatable • Price fluctuation • Unstable supply chain	• Improved soil health and also income • Environment friendly and effective method to break pest and disease cycle • Extra income from maize cultivation	• Problem of sowing in rice crop • Problem in marketability

7

Conclusion

Since the Green Revolution of 1960s, Indian agriculture have been highly dependent on chemical fertilizers, exploitation of ground water for irrigation, intensive tillage etc. to raise the food grain production of the country towards ensuring food security for all. The consequences of these practices resulted in declined and degraded soil health, soil erosion and dried up water capillaries – ultimately posing threat to the sustainability of the future production. This invites the need of affordable changes in the agricultural practices without hampering the production. Conservation Agriculture, as put forward by FAO, is a set of sustainable agricultural practices, aims to conserve, improve and make efficient use of natural resources through management of soil, water and biological resources in aid with external inputs. The three basic principles of Conservation Agriculture include – minimum soil disturbances, permanent soil cover and crop rotation or crop diversification.

To make the maximum benefit out of Conservation Agriculture, one must know how to manage water, nutrient application and soil organic carbon, as these are some critical inputs to agriculture. This requires continuous and rigorous efforts by the extension agents in educating the farmers to raise their consensus in managing the aforesaid resources. The social ecology in which the farmers live, the peers interacting with, plays a vital role in perception development of the farmer. So it is very important to understand the socio-ecological condition of the area as well as the farmers before planning to introduce new practices for its sustainability and longevity. With this background, the present study, "Social Ecology of Conservation Agriculture in terms of Water – Nutrient – Carbon Management: The Analysis and Interpretation" was designed with the following objectives –

i. To study the present status of perception and practices of Conservation Agriculture in terms of water-nutrient-carbon management among the farmers

ii. To identify and standardize the set of variables in operation, impacting on the behavior of water, nutrient and carbon management of the farmers;

iii. To elucidate and interpret the complex interactions amongst and between sets of variables, as predicted and predictor characters, as a system function and policy implication in CA;

iv. To organize series of participatory analyses and action learning to generate a dynamic repository of knowledge covering both indigenous and conventional approaches.

The study was conducted in the state of West Bengal in two agro-ecological zones – New Alluvial zone and Terai zone. From each zone two districts, Nadia and Hooghly from New Alluvial zone and Coochbehar and Alipurduar from Terai zone were selected. Haringhata and Chakdaha Block from Nadia district, Balagarh block from Hooghly district, Coochbehar I, Coochbehar II and Dinhata blocks of Coochbehar district and Falakata and Alipurduar I block of Alipurduar district were purposively selected for conducting the present study. 10 villages from the selected blocks in Nadia, 4 villages from selected blocks in Hooghly, 6 villages from selected blocks in Coochbehar and 4 villages from selected blocks in Alipurduar were purposively selected for the present study. A total of 229 respondents were selected, 75 from New Alluvial Zone and 154 from Terai zone, using non-random snowball sampling method.

The study was conducted with 20 independent variables, viz. age, education, family size, farm size, cropping intensity, number of fragments, annual income, income per capita, income per unit of land, annual expenditure, stubble height, volume of residue, scientific orientation, innovativeness, extension agency contact, information seeking behavior, residue management score, perception on natural resource degradation, number of livestock (cattle) owned and mass media utilization to assess perception of farmers on water management, nutrient management, carbon management and ecosystem management as dependent variable. A proper pre-tested and structured interview schedule and a scale developed with experts' opinion and scale construction method was used to collect relevant information from the respondents. The information collected were then screened, normalized and analyzed using a range of statistical tools including descriptive tools, coefficient of correlation, stepwise regression, path analysis, canonical covariate analysis, cluster analysis and artificial neural network analysis. The analysis were made for both the zones (new alluvial zone and terai zone) in separate and pooled as well.

7.1 Major Findings in New Alluvial zone

- The study reveals the mean age of the respondents were 51.49 years with most having education up to high school. The mean family size of the respondents were 6 members and the mean farm size is 9.73

bighas of land with an average cropping intensity of 260.65 per cent and 7 fragments of land. The average annual income was found to be 239702.00 INR whereas average income per capita and average income per unit of land is 48281.97 INR and 30644.23 INR respectively. The average annual expenditure was 167038.67 INR. Average stubble height maintained by the respondents was 5.46 inches and average residue generation is 7042.61 kgs. Average scientific orientation score was 19.19, innovativeness 14.36, extension agency contact 4.91, information seeking behavior 10.64, residue management score 11.85, perception on natural resource degradation 61.46 and mass media utilization 5.52. On an average, most households owned 2 cattle.

- The results showed that the respondents obtained an average score of 13.40 in perception on water management, 14.01 in perception on nutrient management, 12.11 in perception on carbon management and 13.03 in perception on ecosystem management.
- The results of coefficient of correlation against perception on water management reveals that age of the respondents and stubble height exhibited positive and significant and number of fragments and income per unit of land exhibited negative but significant correlation at 1% level of significance.
- The results of coefficient of correlation against perception on nutrient management reveals that education, stubble height and innovativeness exhibited positive and significant and income per unit of land exhibited negative but significant correlation at 1% level of significance. Information seeking behavior and residue management score was positively and significantly correlated at 5% level of significance.
- The results of coefficient of correlation against perception on carbon management reveals that volume of residue, scientific orientation, information seeking behavior, perception on natural resource degradation and mass media utilization exhibited positive and significant and income per unit of land exhibited negative but significant correlation at 1% level of significance. Age was positively and significantly correlated at 5% level of significance.
- The results of coefficient of correlation against perception on ecosystem management reveals that age of the respondent, stubble height, information seeking behavior and perception on natural resource degradation exhibited positive and significant and income per unit of land exhibited negative but significant correlation at 1% level of significance.

Volume of residue was positively and significantly correlated whereas family size and number of fragments were negatively but significantly correlated at 5% level of significance.

- The path analysis of exogenous variables against perception on water management revealed that stubble height exerted the highest direct effect and income per capita exerted the highest indirect effect. The exogenous variable income per unit of land has routed the highest indirect individual effect on as many as 8 other exogenous variables.
- The path analysis of exogenous variables against perception on nutrient management revealed that innovativeness exerted the highest direct effect and annual expenditure exerted the highest indirect effect. The exogenous variable annual expenditure has routed the highest indirect individual effect on as many as 5 other exogenous variables.
- The path analysis of exogenous variables against perception on carbon management revealed that perception on natural resource degradation exerted the highest direct effect and annual expenditure exerted the highest indirect effect. The exogenous variable perception on natural resource degradation has routed the highest indirect individual effect on as many as 7 other exogenous variables.
- The path analysis of exogenous variables against perception on ecosystem management revealed that income per unit of land exerted the highest direct effect and annual expenditure exerted the highest indirect effect. The exogenous variable income per unit of land has routed the highest indirect individual effect on as many as 9 other exogenous variables.
- The results obtained in stepwise analysis of perception on water management reveal that variables, stubble height, age of the respondents, number of fragments and income per unit of land have been retained at the last step which together explains 51.8 per cent of the variance.
- The results obtained in stepwise analysis of perception on nutrient management reveal that variables, innovativeness, income per unit of land, education, residue management score and stubble height have been retained at the last step which together explains 55.1 per cent of the variance.
- The results obtained in stepwise analysis of perception on carbon management reveal that variables, information seeking behavior, perception on natural resource degradation, volume of residue, mass media utilization, extension agency contact and scientific orientation

have been retained at the last step which together explains 52.6 per cent of the variance.

- The results obtained in stepwise analysis of perception on ecosystem management reveal that variables, income per unit of land, age, stubble height, information seeking behavior and perception on natural resource degradation have been retained at the last step which together explains 49.8 per cent of the variance.
- The canonical covariate analysis reveals the conglomeration of perception on water management, perception on nutrient management and perception on carbon management in the left side that moves along with independent variables age, cropping intensity, annual income, annual expenditure, stubble height, volume of residue, information seeking behavior and number of livestock owned. The other dependent variable, perception on ecosystem management moves alone with the right side variables education, family size, farm size, number of fragments, income per capita, income per unit of land, scientific orientation, innovativeness, extension agency contact, residue management score and mass media utilization.
- The cluster analysis reveals formation of 5 homogenous clusters – cluster 1 with 49 respondents, cluster 2 with 7 respondents, cluster 3 with 12, cluster 4 and cluster 5 with 4 respondents each
- The neural network analysis reveals that out of all the variables passing through hidden layer H1:1 number of fragments has got the highest impact.

7.2 Major Findings of Terai Zone

- The study reveals the mean age of the respondents were 49.43 years with most having education up to middle school. The mean family size of the respondents were 6 members and the mean farm size is 13.29 biogas of land with an average cropping intensity of 237.29 per cent and 2 fragments of land. The average annual income was found to be 179635.71 INR whereas average income per capita and average income per unit of land is 31614.56 INR and 35138.94 INR respectively. The average annual expenditure was 87317.14 INR. Average stubble height maintained by the respondents was 7.57 inches and average residue generation is 8890.64 kgs. Average scientific orientation score was 19.82, innovativeness 15.41, extension agency contact 10.37, information seeking behavior 13.34, residue management score 12.03, perception on natural resource degradation 67.16 and mass media utilization 5.54. On an average, most households owned 4 cattle.

- The results showed that the respondents obtained an average score of 12.25 in perception on water management, 13.38 in perception on nutrient management, 10.71 in perception on carbon management and 12.02 in perception on ecosystem management.
- The results of coefficient of correlation against perception on water management reveals that education, cropping intensity, stubble height, scientific orientation, innovativeness, information seeking behavior and mass media utilization exhibited positive and significant and income per capita exhibited negative but significant correlation at 1% level of significance. Extension agency contact exhibited positive and significant correlation at 5% level of significance.
- The results of coefficient of correlation against perception on nutrient management reveals that age, farm size, volume of residue, residue management score and perception on natural resource degradation exhibited positive and significant correlation at 1% level of significance.
- The results of coefficient of correlation against perception on carbon management reveals that education, cropping intensity, annual income, stubble height, scientific orientation, extension agency contact and mass media utilization exhibited positive and significant correlation at 1% level of significance. Innovativeness was positively and significantly correlated at 5% level of significance.
- The results of coefficient of correlation against perception on ecosystem management reveals that education, cropping intensity, stubble height, scientific orientation, innovativenessand mass media utilization exhibited positive and significantcorrelation at 1% level of significance. Annual income and extension agency contact were positively and significantly correlated income per capita was negatively but significantly correlated at 5% level of significance.
- The path analysis of exogenous variables against perception on water management revealed that stubble height exerted the highest direct effect and mass media utilization exerted the highest indirect effect. The exogenous variable scientific orientation has routed the highest indirect individual effect on as many as 7 other exogenous variables.
- The path analysis of exogenous variables against perception on nutrient management revealed that volume of residue exerted the highest direct

effect and perception on natural resource degradation exerted the highest indirect effect. The exogenous variable farm size has routed the highest indirect individual effect on as many as 10 other exogenous variables.

- The path analysis of exogenous variables against perception on carbon management revealed that education exerted the highest direct effect and scientific orientation exerted the highest indirect effect. The exogenous variable education has routed the highest indirect individual effect on as many as 7 other exogenous variables.
- The path analysis of exogenous variables against perception on ecosystem management revealed that education exerted the highest direct effect and innovativeness exerted the highest indirect effect. The exogenous variable education has routed the highest indirect individual effect on as many as 5 other exogenous variables.
- The results obtained in stepwise analysis of perception on water management reveal that variables, innovativeness, scientific orientation, stubble height, cropping intensity, education and extension agency contact have been retained at the last step which together explains 51.6per cent of the variance.
- The results obtained in stepwise analysis of perception on nutrient management reveal that variables, volume of residue, farm size, age, perception on natural resource degradation, family size and residue management score have been retained at the last step which together explains 49.2 per cent of the variance.
- The results obtained in stepwise analysis of perception on carbon management reveal that variables, mass media utilization, education, stubble height, scientific orientation, extension agency contact, number of livestock (cattle) owned, and cropping intensity have been retained at the last step which together explains 48.9 per cent of the variance.
- The results obtained in stepwise analysis of perception on ecosystem management reveal that variables, scientific orientation, stubble height, education, mass media utilization, innovativeness, cropping intensity, number of livestock owned (cattle) and age have been retained at the last step which together explains 52.7 per cent of the variance.
- The canonical covariate analysis reveals the conglomeration of perception on water management and perception on carbon management in the left side that moves along with independent variables farm size, number of fragments, annual income, income per capita, stubble height

and volume of residue. The other dependent variable, perception on ecosystem management and perception on nutrient management moves together with the right side variables age, education, family size, cropping intensity, income per capita, annual expenditure, scientific orientation, innovativeness, extension agency contact, information seeking behavior, residue management score, perception on natural resource degradation, number of livestock owned and mass media utilization.

- The cluster analysis reveals formation of 4 homogenous clusters – cluster 1 with 118 respondents, cluster 2 with 9 respondents, cluster 3 with 19 and cluster 4 with 8 respondents each
- The neural network analysis reveals that out of all the variables passing through hidden layer H1:1 age, education and residue management score has got the highest impact.

7.3 Major Findings of Pooled (New Alluvial and Terai zone) Analysis

- The study reveals the mean age of the respondents were 50.10 years with most having education up to middle school. The mean family size of the respondents were 6 members and the mean farm size is 12.12 bighas of land with an average cropping intensity of 244.94 per cent and 4 fragments of land. The average annual income was found to be 199308.08 INR whereas average income per capita and average income per unit of land is 37073.32 INR and 33666.87 INR respectively. The average annual expenditure was 113426.81 INR. Average stubble height maintained by the respondents was 6.88 inches and average residue generation is 8285.39 kgs. Average scientific orientation score was 19.61, innovativeness 15.07, extension agency contact 8.58, information seeking behavior 12.45, residue management score 11.97, perception on natural resource degradation 65.30 and mass media utilization 5.54. On an average, most households owned 3 cattle.
- The results showed that the respondents obtained an average score of 12.63 in perception on water management, 13.59 in perception on nutrient management, 11.17 in perception on carbon management and 12.35 in perception on ecosystem management.
- The results of coefficient of correlation against perception on water management reveals that age, education, cropping intensity, stubble height, scientific orientation and innovativeness exhibited positive and significant correlation at 1% level of significance. Mass media utilization exhibited positive and number of livestock owned has negative but significant correlation at 5% level of significance.

- The results of coefficient of correlation against perception on nutrient management reveals that age, education, farm size, annual income, volume of residue, residue management score and perception on natural resource degradation exhibited positive and significant correlation at 1% level of significance. Income per capita has positive and significant correlation at 5% level of significance
- The results of coefficient of correlation against perception on carbon management reveals that education, cropping intensity, number of fragments, annual income, stubble height, scientific orientation and mass media utilization exhibited positive and significant whereas number of livestock has exhibited a negative and significantcorrelation at 1% level of significance.
- The results of coefficient of correlation against perception on ecosystem management reveals that education, cropping intensity, annual income, stubble height, scientific orientation, innovativeness and mass media utilization exhibited positive and significant and number of livestock exhibited a negative but significant correlation at 1% level of significance. Age was positively and significantly correlated and family size was negatively but significantly correlated at 5% level of significance.
- The path analysis of exogenous variables against perception on water management revealed that age of the respondent has exerted the highest direct effect and innovativeness exerted the highest indirect effect. The exogenous variable annual expenditure has routed the highest indirect individual effect on as many as 6 other exogenous variables.
- The path analysis of exogenous variables against perception on nutrient management revealed that volume of residue exerted the highest direct effect and income per capita exerted the highest indirect effect. The exogenous variable volume of residue has routed the highest indirect individual effect on as many as 7 other exogenous variables.
- The path analysis of exogenous variables against perception on carbon management revealed that mass media utilization exerted the highest direct effect and innovativeness exerted the highest indirect effect. The exogenous variable number of fragments has routed the highest indirect individual effect on as many as 6 other exogenous variables.
- The path analysis of exogenous variables against perception on ecosystem management revealed that education exerted the highest direct effect and innovativeness exerted the highest indirect effect. The

exogenous variable annual expenditure has routed the highest indirect individual effect on as many as 6 other exogenous variables.

- The results obtained in stepwise analysis of perception on water management reveal that variables, stubble height, education, age, cropping intensity, scientific orientation, innovativeness, number of livestock owned and income per unit of land have been retained at the last step which together explains 36.7 per cent of the variance.
- The results obtained in stepwise analysis of perception on nutrient management reveal that variables, volume of residue, age, residue management score, farm size, extension agency contact, family size, perception on natural resource degradation, and number of livestock owned have been retained at the last step which together explains 41.1 per cent of the variance.
- The results obtained in stepwise analysis of perception on carbon management reveal that variables, mass media utilization, education, scientific orientation, number of livestock (cattle) owned, number of fragments, perception on natural resource degradation, cropping intensity and age have been retained at the last step which together explains 38.8 per cent of the variance.
- The results obtained in stepwise analysis of perception on ecosystem management reveal that variables, scientific orientation, education, mass media utilization, age, cropping intensity, number of livestock owned (cattle), stubble height, income per unit of land, family size and innovativeness have been retained at the last step which together explains 46.3 per cent of the variance.
- The canonical covariate analysis reveals the conglomeration of perception on water management and perception on carbon management in the left side that moves along with independent variables farm size, annual income, annual expenditure, stubble height and volume of residue. The other dependent variable, perception on ecosystem management and perception on nutrient management moves together with the right side variables age, education, family size, cropping intensity, number of fragments, income per capita, income per unit of land, scientific orientation, innovativeness, extension agency contact, information seeking behavior, residue management score, residue management score, perception on natural resource degradation, number of livestock owned and mass media utilization.

- The cluster analysis reveals formation of 5 homogenous clusters – cluster 1 with 203 respondents, cluster 2 with 7 respondents, cluster 3 with 12 and cluster 4 with 4 respondents and cluster 5 with 3 respondents.
- The neural network analysis reveals that out of all the variables passing through hidden layer H1:1 age, volume of residue, innovativeness and residue management score has got the highest impact.

7.4 Conclusion

The prime concern of Indian agriculture is the faster and robust deterioration of its ecological base and resilience. Every time it is adding higher entropy and anomy both to its agro-ecosystem and social ecology. When ecological foundation of agricultural production process turns fragile, the production and productivity both will go unpredictable and undulating. The narrowing down of gene bases, erosion of top soils, decline of biodiversity are clinically linked to poverty, malnutrition and socio-political unrest as well. Agriculture cannot be a proposition for decades rather it has to go for a marathon race covering centuries. Besides, the societal growth, political balances, food and nutritional security, entitlement and empowerment and environmental sanctity as well as ecological balances are to be assured by stable, dedicated and evolving agriculture. In doing so, CA is the magnum step to both reenergize and restore ecological balances and services. The present study covers two conspicuous agro-climatic zones of West Bengal viz. New Alluvial Zone and Terai Zone.

The four predicted components of perception on water management, perception on nutrient management, perception on carbon management and perception on ecosystem management which have been estimated through a score of 20 socio-economic, techno-managerial and socio-ecological variables. The empirical results have provided the pre-cursors for conclusion for both the zones. For alluvial zone, it is well discernable that unless issues of income, precisely to mention net profits from CA, is being assured, it would be very difficult for CA to survive Similarly, stubble burning has a classical problem for this zone and again unless the stubbles are retained in the field and blended with the soil, the soil health as well as organic carbon for this zone would be very difficult to maintain. Due to humongous population pressure and disintegration of joint families into single ones have contributed to the faster fragmentations of holdings which again have contributed in turning the tiny farms both energy and resource prodigal. So it can be concluded that unless a community based CA approach is being accultured and accommodated into the farming behavior, the basics principles of CA would be difficult to achieve.

For terai zone, which is already blessed with bounties of rainfall and riverine networks, the perception of soil and water conservation has been difficult to imprint into the mind and action of farmers. But this zone has got a profile of CA experience for the couple of decades and institutional supports from FPOs or SAU (UBKV) have been forthcoming to make the adaptation of CA. The young generations are increasingly being involve in socialization of CA. Unlike the southern counterpart, i.e. Alluvial zone, here CA has become a profitable venture. The area under maize has constantly been increasing to replace classical rice and jute crops. The dominant variables which are operating in adoption of CA have been communication network, entrepreneurship, livestock count and income. So, more of entrepreneurial training, market segmentations and intelligence, formation of common interest groups can be successful in this zone.

Both the zones of west Bengal has again offered a stark difference from northern part of India, and this is simply because while Punjab and Haryana has got 600-700 annual rainfall, this part of WB has got 1600mm rainfall for alluvial and 3000mm rainfall for terai zone. So in general rainwater harvesting, mulching of soils, minimal disruption of soils, upscaling of crop rotation are more important. Through rainwater harvesting, we can retrench GW extraction, which again is contaminated with toxic heavy metals.

The crux of the issue would be to make CA a dependable means and ways of livelihood for millions of farmers in India. When farmers start getting *happy returns* from CA, the soil, water and biota of the surrounding ecosystem will start giggling, and the good earth will turn again a happy abode for millions and millions of children and of with security of food and happiness of livelihood. What we need is a strong policy support, environmental laws and community mobilization to usher a social movement for the good of the mankind. The fragmenting land holding is posing higher threats than its tiny average size, associated with it are a wrong feat of agricultural modernization. The non-cognitive adoption of conservation agriculture is a silver line no doubt, but its transformation into a conscious and concerted socialization is more and more important. The statistically vindicated relation shows that unless conservation agriculture is economically viable and socially convincing, its success will be remote, else, we can opt for a dedicated and well directed mobilization for conservation agriculture as an ecological rhetoric and economical promise.

8

Recommendation, Limitation and Future Scope of The Study

8.1 Recommendations

The specific recommendation for NAZ in promoting CA are

- Since fragmentation has come out as a barrier to CA, the community based CA entrepreneurship will be more successful to make CA ecologically resilient, economically viable and socially adoptable.
- Crop residues are viable source for soil nutrients including organic carbon replenishment, but unfortunately left over crop residues are either burnt out or wrongly used. So, a community bank on crop residues in the form of bio-parks on a commercial basis can be a dependable source of organic manure for being blended into the farm soil. Auditing on volume of disposable CR and its proper application can help in making operating farm more resilient and productive.
- Since CA farmers have to go through a minimum gestation period and they are expected to contribute to the ecological resilience and carbon sequestration, economic incentives must be extended for their productive as well as stewardship function. A policy promulgation is the need of the hour to support their contribution
- Since fragmentation of holding poses a potential threat to resource energy biodiversity and economic conservation, so a strong land and farm policy needs to be there to abate disintegration and fragmentation of holding
- CA based entrepreneurship can go a long way in offering both economic and social sustenance to the project and the concept as well. CA should be considered not as a bunch of techniques but as an integration of concepts and operation to save the ecosystem from further degradation and denudation.

8.2 Limitations

1. Having been constrained by pandemic situation another important climatic zone i.e. coastal ecosystem could not be included in the study. Only some few farmers could have been interrogated
2. In the sampling process snowballing process or contacting farmers through cross referencing has been followed. Else, it could have been through an ideal systematic random sampling
3. Since the farmers are mostly following conservation agricultural practices, especially in new alluvial zone, through non-cognitive adoption or inductive adoption, it has been very difficult to depict a normal distribution of farmers adopting conservation agriculture through a conscious or logical adoption
4. Conservation agriculture is reaching the farmers just as a rhetoric. That is why conscious and active adoption of CA has not been elucidated
5. The gender representation in the entire socialization of conservation agriculture has not been that conspicuous that's why a gender depiction in this research is missing grossly. However future research is expected to take care of it.
6. More advanced software on data analytics and deep learning could have been effective in predicting the future scenario

8.3 Future Scope of Research

- Since the progress of CA remains at an initial stage in India rather, huge scope of CA research are there to spare head the seamless generation of CA techniques and adaptability. While globally 180mha of land under CA, but India's contribution is as low as 1.5mha. So a constraint and content analysis of CA in terms of its future progress can itself be a text of research.
- The scope and opportunities for application of data analytics in the domains of visualization, prediction and prescription, can offer a new genre of CA research
- Customization, cataloguing and modelling of CA research itself merits a new plank of research.
- The social ecology of conservation research including energy and social metabolism, water and biodiversity stewardship, entries ecology and frame working, gender, empowerment through participatory resource monitoring are of huge research interest and application

- A seamless research for CA auditing, forecasting, monitoring, tracking can provide huge theoretical and empirical construct for spare heading CA research in future and posterity.

References

Acharya, S. K., & Chatterjee, R. (2019). Conservation Agriculture: The dynamics of Ecology and Ecological Services. Call for Editorial Board Members, 5(2), 69.

Adger, W. N. (2000). Social and ecological resilience: are they related? Progress in human geography, 24(3), 347-364.

Ahmed, S. A., Karablieh, E. K., & Al-Kadi, A. S. (2004). An investigation into the perceived farm management and marketing educational needs of farm operations in Jordan. Journal of Agricultural Education, 45(3), 34-43.

Andersson, J. A., & D'Souza, S. (2014). From adoption claims to understanding farmers and contexts: A literature review of Conservation Agriculture (CA) adoption among smallholder farmers in southern Africa. Agriculture, ecosystems & environment, 187, 116-132.

Arfanuzzaman, M., & Rahman, A. A. (2017). Sustainable water demand management in the face of rapid urbanization and ground water depletion for social–ecological resilience building. Global Ecology and Conservation, 10, 9-22.

Bacon, C. M., Getz, C., Kraus, S., Montenegro, M., & Holland, K. (2012). The social dimensions of sustainability and change in diversified farming systems. Ecology and Society, 17(4).

Badgujjar, M. K. (2012) A Study on knowledge and adoption of organic farming practices among the farmers in Sehore district (M.P.). Unpublished M. Sc. Thesis. Rajmata Vijayaraje Scindia Krishi Vishwa Vidyalaya, Gwalior.

Ballard, H. L., & Belsky, J. M. (2010). Participatory action research and environmental learning: implications for resilient forests and communities. Environmental Education Research, 16(5-6), 611-627.

Barber, R., Englisch, G., Douglas, M., Nabhan, H., Bot, A., Roy, R., ... & Benites, J. (2000). Guidelines and reference material on integrated soil and nutrient management and conservationfor farmer field schools.

Belay, S. A., Assefa, T. T., Prasad, P. V., Schmitter, P., Worqlul, A. W., Steenhuis, T. S., ... & Tilahun, S. A. (2020). The response of water and nutrient dynamics and of crop yield to conservation agriculture in the Ethiopian highlands. Sustainability, 12(15), 5989.

Bharti, C., Mohapatra, A., Maurya, R., Maharana, C., Maurya, A., & Malakar, P. (2020). Nitrate pollution with modernization of Indian agriculture. Journal of Pharmacognosy and Phytochemistry, 9(3), 2073-2080.

Bruening, H. T., Radhakrishma, R. B. & Rollins, T. J. (1992) Environmental issues: farmers' perception about usefulness of informational and organizational sources. Journal of Agricultural Education, 33(2), 34-42.

Busari, M. A., Kukal, S. S., Kaur, A., Bhatt, R., & Dulazi, A. A. (2015). Conservation tillage impacts on soil, crop and the environment. International soil and water conservation research, 3(2), 119-129.

Cary, J. W., & Wilkinson, R. L. (1997). Perceived profitability and farmers 'conservation behaviour. Journal of agricultural economics, 48(1□3), 13-21.

Chapin III, F. S., Carpenter, S. R., Kofinas, G. P., Folke, C., Abel, N., Clark, W. C., ... & Swanson, F. J. (2010). Ecosystem stewardship: sustainability strategies for a rapidly changingplanet. Trends in ecology & evolution, 25(4), 241-249.

Chapin III, F. S., Sommerkorn, M., Robards, M. D., & Hillmer-Pegram, K. (2015). Ecosystem stewardship: A resilience framework for arctic conservation. Global Environmental Change, 34, 207-217.

Chisanga, K., Kafwamfwa, N., Hamazakaza, P., Mwila, M., Sinyangwe, J., & Lungu, O. (2017). Farmer Perceptions of Conservation Agriculture in Maize-Legume Systems for Small- Holder Farmers in Sub Saharan Africa-A Beneficiary Perspective in Zambia. International journal of Horticulture, Agriculture and Food science (IJHAF), 1(3), 10-15.

Corsi, S., & Muminjanov, H. (2019). Conservation Agriculture: Training guide for extension agents and farmers in Eastern Europe and Central Asia. Food and Agriculture Organization United Nations.

Cross, M. S., McCarthy, P. D., Garfin, G., Gori, D., & Enquist, C. A. (2013). Accelerating adaptation of natural resource management to address climate change. Conservation Biology, 27(1), 4-13.

Cumming, G. S., Barnes, G., Perz, S., Schmink, M., Sieving, K. E., Southworth, J., ... & Van Holt, T. (2005). An exploratory framework for the empirical measurement of resilience. Ecosystems, 8(8), 975-987.

Daxini, A., O'Donoghue, C., Ryan, M., Buckley, C., Barnes, A. P., & Daly, K. (2018). Which factors influence farmers' intentions to adopt nutrient management planning?. Journal of environmental management, 224, 350-360.

Di Falco, S., & Chavas, J. P. (2008). Rainfall shocks, resilience, and the effects of crop biodiversity on agroecosystem productivity. Land Economics, 84(1), 83-96.

Donovan, M. (2020). CIMMYT News. https://www.cimmyt.org/news/what-is-conservation-agriculture

Du, L., McCarty, G. W., Li, X., Rabenhorst, M. C., Wang, Q., Lee, S., ... & Zou, Z. (2021). Spatial extrapolation of topographic models for mapping soil organic carbon using local samples. Geoderma, 404, 115290.

Dumanski, J., Peiretti, R., Benites, J. R., McGarry, D., & Pieri, C. (2006). The paradigm of conservation agriculture. Proceedings of world association of soil and water conservation, 1(2006), 58-64.

Duncan, D. W. (2004). Knowledge and perceptions of Virginia secondary agriculture educators toward the agricultural technology program at Virginia Tech. Journal of Agricultural Education, 45(1), 21-28.

Dutta, A. 2012. A study on practices and perception of farmers towards farm diversification in Burdwan district of West Bengal. Unpublished Thesis, M. Sc. G. B. Pant University of Agriculture and Technology (GBPUA&T), Pantnagar, Uttarakhand.

Enfors, E. (2009). Traps and transformations: Exploring the potential of water system innovations in dryland sub-Saharan Africa (Doctoral dissertation, Department of Systems Ecology, Stockholm University).

Ervin, C. A., & Ervin, D. E. (1982). Factors affecting the use of soil conservation practices: hypotheses, evidence, and policy implications. Land economics, 58(3), 277-292.

FAO. (2013a). Conservation Agriculture in Central Asia: Status, Policy, Institutional Support, and Strategic Framework for Its Promotion.

FAO. (2013b). Policy and Institutional support for conservation agriculture in the Asia-Pacific Region.

Farooq, M., & Siddique, K. H. (2015). Conservation agriculture: concepts, brief history, and impacts on agricultural systems. In Conservation agriculture (pp. 3-17). Springer, Cham.

Farooq, M., Flower, K. C., Jabran, K., Wahid, A., & Siddique, K. H. (2011). Crop yield and weed management in rainfed conservation agriculture. Soil and tillage research, 117, 172-183.

Farouque, M. (2007). Farmers' Perception of Integrated Soil Fertility and Nutrient Management for Sustainable Crop Production: A Study of Rural Areas in Bangladesh. Journal of Agricultural Education, 48(3), 111-122.

Fiksel, J. (2006). Sustainability and resilience: toward a systems approach. Sustainability: Science, Practice and Policy, 2(2), 14-21.

Flores, B. M., Staal, A., Jakovac, C. C., Hirota, M., Holmgren, M., & Oliveira, R. S. (2020). Soil erosion as a resilience drain in disturbed tropical forests. Plant and Soil, 450(1), 11-25.

Foster, D., Swanson, F., Aber, J., Burke, I., Brokaw, N., Tilman, D., & Knapp, A. (2003). The importance of land-use legacies to ecology and conservation. BioScience, 53(1), 77-88.

Friedrich, T., Derpsch, R., & Kassam, A. (2012). Global overview of the spread of conservation agriculture. Field Actions Sci, Rep 6:1-7.

Garbach, K., Milder, J. C., DeClerck, F. A., Montenegro de Wit, M., Driscoll, L., & Gemmill-Herren, B. (2017). Examining multi-functionality for crop yield and ecosystem services in fivesystems of agroecological intensification. International Journal of Agricultural Sustainability, 15(1), 11-28.

Gholve, M. A. (1986). Socio-cultural and psychological correlates of tribal development in Kinwat block of Nanded district of Maharashtra state (Doctoral dissertation, Vasantrao Naik Marathwada Krishi Vidyapeeth, Parbhani).

Ghosh, S., Das, T. K., Sharma, D., & Gupta, K. (2019). Potential of conservation agriculture for ecosystem services: A review. Indian Journal of Agricultural Sciences, 89(10), 1572-1579.

Giller, K. E., Andersson, J. A., Corbeels, M., Kirkegaard, J., Mortensen, D., Erenstein, O., & Vanlauwe, B. (2015). Beyond conservation agriculture. Frontiers in plant science, 6, 870.

Giller, K. E., Witter, E., Corbeels, M., & Tittonell, P. (2009). Conservation agriculture and smallholder farming in Africa: the heretics' view. Field crops research, 114(1), 23-34.

Greiner, R., & Gregg, D. (2011). Farmers' intrinsic motivations, barriers to the adoption of conservation practices and effectiveness of policy instruments: Empirical evidence from northern Australia. Land use policy, 28(1), 257-265.

Grönwall, J., & Oduro-Kwarteng, S. (2018). Groundwater as a strategic resource for improved resilience: a case study from peri-urban Accra. Environmental Earth Sciences, 77(1), 6.

Haggblade, S., & Tembo, G. (2003). Conservation farming in Zambia. Intl Food Policy Res Inst.

Hails, R. S., & Ormerod, S. J. (2013). Ecological science for ecosystem services and the stewardship of N atural C apital.

Hobbs, P. R., Sayre, K., & Gupta, R. (2008). The role of conservation agriculture in sustainable agriculture. Philosophical Transactions of the Royal Society B: Biological Sciences, 363(1491), 543-555.

Holling, C. S. (1973). Resilience and stability of ecological systems. Annual review of ecology and systematics, 4(1), 1-23.

Holling, C. S. (1986). The resilience of terrestrial ecosystems: local surprise and global change. Sustainable development of the biosphere, 14, 292-317.

Hurni, H., Herweg, K., Portner, B., & Liniger, H. (2008). Soil erosion and conservation in global agriculture. In Land use and soil resources (pp. 41-71). Springer, Dordrecht.

J. M. (2015). Biodiversity and resilience of ecosystem functions. Trends in ecology & evolution, 30(11), 673-684.

Jabbar, M. A., Saleem, M. M., Gebreselassie, S., & Beyene, H. (2003). Role of knowledge inthe adoption of new agricultural technologies: an approach and an application. International Journal of Agricultural Resources, Governance and Ecology, 2(3-4), 312-327.

Jackson, L., van Noordwijk, M., Bengtsson, J., Foster, W., Lipper, L., Pulleman, M., ... & Vodouhe, R. (2010). Biodiversity and agricultural sustainability: from assessment to adaptivemanagement. Current opinion in environmental sustainability, 2(1-2), 80-87.

Johansen, C., Haque, M. E., Bell, R. W., Thierfelder, C., & Esdaile, R. J. (2012). Conservation agriculture for small holder rainfed farming: Opportunities and constraints of new mechanizedseeding systems. Field Crops Research, 132, 18-32.

Jost, C., Kyazze, F., Naab, J., Neelormi, S., Kinyangi, J., Zougmore, R., ... & Kristjanson, P. (2016). Understanding gender dimensions of agriculture and climate change in smallholder farming communities. Climate and Development, 8(2), 133-144.

Kanavi, V.P. 2000. Study on knowledge and adoption behaviour of sugarcane growers in Belgaum district of Karnataka. Unpublished Thesis, M. Sc. (Ag). University of Agricultural Sciences, Dharwad, Karnataka. 20p.

Kashem, M. A., & Mikuni, H. (1998). Towards sustainable agricultural development: Use of indigenous technical knowledge (ITK) by the farmers of Bangladesh and Japan. Research Monograph, 2.

Kassam, A., Friedrich, T., & Derpsch, R. (2019). Global spread of conservationagriculture. International Journal of Environmental Studies, 76(1), 29-51.

Kassam, A., Friedrich, T., Shaxson, F., Bartz, H., Mello, I., Kienzle, J., & Pretty, J. (2014). Field actions science reports. Field Actions Science Reports, 7, 1-12.

Katic, P., & Grafton, R. Q. (2011). Optimal groundwater extraction under uncertainty:resilience versus economic payoffs. Journal of Hydrology, 406(3-4), 215-224.

King, C. A. (2008). Community resilience and contemporary agri□ecological systems: reconnecting people and food, and people with people. Systems Research and Behavioral Science: The Official Journal of the International Federation for Systems Research, 25(1), 111-124.

Knickel, K. (1990). Agricultural structural change: Impact on the rural environment. Journal of rural studies, 6(4), 383-393.

Knowler, D., & Bradshaw, B. (2007). Farmers' adoption of conservation agriculture: A review and synthesis of recent research. Food policy, 32(1), 25-48.

Kosmas, C., Karamesouti, M., Kounalaki, K., Detsis, V., Vassiliou, P., & Salvati, L. (2016). Land degradation and long-term changes in agro-pastoral systems: An empirical analysis of ecological resilience in Asteroussia-Crete (Greece). Catena, 147, 196-204.

Kumaraswamy, S., & Kunte, K. (2013). Integrating biodiversity and conservation with modern agricultural landscapes. Biodiversity and conservation, 22(12), 2735-2750.

Lal, R. (1997) Degradation and resilience of soils. Philosophical Transactions of the Royal Society of London. Series B: Biological Sciences, 352(1356), 997-1010.

Lal, R. (2015). Sequestering carbon and increasing productivity by conservation agriculture. Journal of Soil and Water Conservation, 70(3), 55A-62A.

Lopes, A. A., Viriyavipart, A., & Tasneem, D. (2020). The role of social influence in crop residue management: Evidence from Northern India. Ecological Economics, 169, 106563.

Lugandu, S. (2013, April). Factors influencing the adoption of conservation agriculture by smallholder farmers in Karatu and Kongwa districts of Tanzania. In REPOA's18thannual research workshop held at the Kunduchi Beach Hotel, Dar es Salaam, Tanzania.

M. (2008). Intensification of New Zealand agriculture: implications for biodiversity. New Zealand Journal of Agricultural Research, 51(3), 253-263.

Maggioni, E. (2015). Water demand management in times of drought: What matters for water conservation. Water Resources Research, 51(1), 125-139.

Mahenge, J. (2014). Comparative economic analysis of conservation and conventional agricultural practices in Southern Uluguru Mountains, Morogoro, Tanzania (Doctoral dissertation, Sokoine University of Agriculture).

Mathevet, R., Bousquet, F., Larrère, C., & Larrère, R. (2018). Environmental stewardship and ecological solidarity: rethinking social-ecological interdependency and responsibility. Journalof Agricultural and Environmental Ethics, 31(5), 605-623

McNeely, J. A., & Schroth, G. (2006). Agroforestry and biodiversity conservation–traditional practices, present dynamics, and lessons for the future. Biodiversity & Conservation, 15(2), 549-554.

Meena, D. S. 2012. A study on knowledge level of farmers of Farmers' Field School (FFS) regarding Integrated Crop Management (ICM) practices in district Udham Singh Nagar of Uttarakhand. Unpublished Thesis, M. Sc. G. B. Pant University of Agriculture and Technology(GBPUA&T), Pantnagar, Uttarakhand.

Mohapatra, L. (2013). A study on adoption of soil health management practices by farmers in Mayurbhanj district of Odisha (Doctoral dissertation, GB Pant University of Agriculture and Technology, Pantnagar-263145 (Uttarakhand)).

Moller, H., MacLeod, C. J., Haggerty, J., Rosin, C., Blackwell, G., Perley, C., ... & Gradwohl,

Moore, E. G. (1988). Perception of teacher educators in agriculture relating to agricultural and rural improvements in developing countries. Journal of the American Association of Teacher Educators in Agriculture, 29(2), 7-13.

Mugandani, R., & Mafongoya, P. (2019). Behaviour of smallholder farmers towards adoption of conservation agriculture in Zimbabwe. Soil Use and Management, 35(4), 561-575.

Mwase, W., Sefasi, A., Njoloma, J., Nyoka, B. I., Manduwa, D., & Nyaika, J. (2015). Factors affecting adoption of agroforestry and evergreen agriculture in Southern Africa. Environmentand Natural Resources Research, 5(2), 148.

Naik, A., Sreenivasulu, M., Sreenivasa Rao, I. & Lankati, M. (2018). A Study on Knowledge Level of Farmers on
Organic Red Gram Cultivation Practices in Dryland Areas of Karnataka, India. International Journal of Current Microbiologyand Applied Sciences. 7 (3): 435-440.

Nearing, M. A., Xie, Y., Liu, B., & Ye, Y. (2017). Natural and anthropogenic rates of soil erosion. International Soil and Water Conservation Research, 5(2), 77-84.

Nelson, E., Mendoza, G., Regetz, J., Polasky, S., Tallis, H., Cameron, D., ... & Shaw, M. (2009). Modeling multiple ecosystem services, biodiversity conservation, commodity production, and tradeoffs at landscape scales. Frontiers in Ecology and the Environment, 7(1),4-11.

Nichols, J. D., Koneff, M. D., Heglund, P. J., Knutson, M. G., Seamans, M. E., Lyons, J. E., & Williams, B. K. (2011). Climate change, uncertainty, and natural resource management. TheJournal of Wildlife Management, 75(1), 6-18.

Nigeria. Journal of Organic Systems, 6(1).

Ntshangase, N. L., Muroyiwa, B., & Sibanda, M. (2018). Farmers' perceptions and factors influencing the adoption of no-till conservation agriculture by small-scale farmers in Zashuke,KwaZulu-Natal Province. Sustainability, 10(2), 555.

Nyambose, W., & Jumbe, C. B. (2013). Does conservation agriculture enhance household food security? Evidence from smallholder farmers in Nkhotakota in Malawi (No. 309-2016-5147).

Odum, H. T. (1996). Enviromental accounting: emergy and enviromental decision making (No. 657.73 O-27e). New York, US: Wiley.

Oliver, T. H., Heard, M. S., Isaac, N. J., Roy, D. B., Procter, D., Eigenbrod, F., ... & Bullock,

Oyesola, O. B., & Obabire, I. E. (2011). Farmers' Perceptions of Organic Farming In Selected Local Government Areas Of Ekiti State,

Palm, C., Blanco-Canqui, H., DeClerck, F., Gatere, L., & Grace, P. (2014). Conservation agriculture and ecosystem services: An overview. Agriculture, Ecosystems & Environment, 187, 87-105.

Pandey, D. (2016). Conservation Agriculture: why is adoption not increasing as expected?. Agrilinks

Patel, K. P. (2013). Development of scale to measure attiitude of farmers towards green manuriing for sustaiinable agriiculture (Doctoral dissertation, AAU, Anand).

Pellegrini, L., & Tasciotti, L. (2014). Crop diversification, dietary diversity and agricultural income: empirical evidence from eight developing countries. Canadian Journal of Development Studies/Revue canadienne d'études du développement, 35(2), 211-227.

Perry-Hill, R., & Prokopy, L. S. (2014). Comparing different types of rural landowners: Implications for conservation practice adoption. Journal of Soil and Water Conservation, 69(3), 266-278.

Pillay, N. K., & Maharaj, P. (2013). Population ageing in Africa. In Aging and health in Africa (pp. 11-51). Springer, Boston, MA.

Pittelkow, C. M., Liang, X., Linquist, B. A., Van Groenigen, K. J., Lee, J., Lundy, M. E., & Van Kessel, C. (2015). Productivity limits and potentials of the principles of conservation agriculture. Nature, 517(7534), 365-368.

Power, A. G. (2010). Ecosystem services and agriculture: tradeoffs andsynergies. Philosophical transactions of the royal society B: biological sciences, 365(1554), 2959-2971.

Pramanik, P., Sharma, D. K. and Maity, A. (2014). Environmental benefits of Conservation Agriculture. Indian Farming 64(8):26–30

Raghunandan, H. C. (2004). A study on knowledge and adoption level of soil and water conservation practices by farmers in northern Karnataka (Doctoral dissertation, University ofAgricultural Sciences GKVK, Bangalore).

Ramamurthy, H. S. (1976). Some factors influencing IR-8 adopter's perception of innovations and their relation to adoption. Unpublished doctoral dissertation, Haryana Agricultural University, India.

Rayangoudar, R. (2009). Knowledge of rural women about organic farming (Doctoral dissertation, UAS, Dharwad).

Ruiz-Colmenero, M., Bienes, R., Eldridge, D. J., & Marques, M. J. (2013). Vegetation cover reduces erosion and enhances soil organic carbon in a vineyard in the central Spain. Catena, 104, 153-160.

Sakharkar Vilas, S. A study on knowledge, fertilizer use pattern and constraints in thecultivation of soybean by farmers of nagpur district, maharashtra (Doctoral dissertation, University of Agricultural Science, Bangalore).

Sanderson, M. A., Archer, D., Hendrickson, J., Kronberg, S., Liebig, M., Nichols, K., ... & Aguilar, J. (2013). Diversification and ecosystem services for conservation agriculture: outcomes from pastures and integrated crop–livestock systems. Renewable agriculture and food systems, 28(2), 129-144.

Sheikh, A. D., Rehman, T., & Yates, C. M. (2003). Logit models for identifying the factors that influence the uptake of new 'no-tillage'technologies by farmers in the rice–wheat and thecotton–wheat farming systems of Pakistan's Punjab. Agricultural systems, 75(1), 79-95.

Shortle, J. S., & Miranowski, J. A. (1986). Effects of risk perceptions and other characteristicsof farmers and farm operations on the adoption of conservation tillage practices. Applied Agricultural Research, 1(2), 85-90.

Singh, S. P., Singh, J., Tyagi, P. K., & Kumar, R. (2010). Relative Influence of Socio-Personal Characteristics and Utilization of Fertilizer Technology in Wheat Cultivation. Indian Journalof Extension Education, 46(1,2), 112-115.

Srinivasarao, C., Lal, R., Kundu, S., & Thakur, P. B. (2015). Conservation agriculture and soil carbon sequestration. Conservation agriculture, 479-524.

Su, Y., Gabrielle, B., & Makowski, D. (2021). A global dataset for crop production under conventional tillage and no tillage systems. Scientific data, 8(1), 1-17.

Sundstrom, S. M., & Allen, C. R. (2019). The adaptive cycle: More than a metaphor. Ecological Complexity, 39, 100767.

Suryawanshi, A., Dubey, M. K., & Saryam, M. Study on the Socio-Economic Profile And Knowledge Level About Vermicompost Technology Among Trainees. Young (Up to 35 years), 26, 22-41.

Tapoopi, M., Kamwi, J. M., & Siyambango, N. (2018). Perception of farmers on conservation agriculture for climate change adaptation in Namibia.

Tatlıdil, F. F., Boz, İ., & Tatlidil, H. (2009). Farmers' perception of sustainable agriculture and its determinants: a case study in Kahramanmaras province of Turkey. Environment, development and sustainability, 11(6), 1091-1106.

Teague, W. R., Apfelbaum, S., Lal, R., Kreuter, U. P., Rowntree, J., Davies, C. A., & Byck, P. (2016). The role of ruminants in reducing agriculture's carbon footprint in North America. Journal of Soil and Water Conservation, 71(2), 156-164.

Thierfelder, C., & Wall, P. C. (2010). Rotation in conservation agriculture systems of Zambia: effects on soil quality and water relations. Experimental agriculture, 46(3), 309-325.

Thiranjangowda, B.T. 2005. A study on knowledge and adoption of soil and water conservation practices by farmers in north Karnataka. M. Sc. Thesis. University of Agricultural Science, Dharwad, Karnataka.

Whitlow, R. (1988). Soil erosion and conservation policy in Zimbabwe: past, present and future. Land Use Policy, 5(4), 419-433.

Wilkinson, C. (2012). Social-ecological resilience: Insights and issues for planning theory. Planning theory, 11(2), 148-169.

Williams, D. L., & Wise, K. L. (1997). Perceptions of Iowa secondary school agricultural education teachers and students regarding sustainable agriculture. Journal of Agricultural Education, 38, 15-20.

Winn, M. I., & Pogutz, S. (2013). Business, ecosystems, and biodiversity: New horizons for management research. Organization & Environment, 26(2), 203-229.

Worrell, R., & Appleby, M. C. (2000). Stewardship of natural resources: definition, ethical and practical aspects. Journal of agricultural and environmental ethics, 12(3), 263-277.

Index

N

O

P

Q

R

S

T